Patent Pending

Patent Pending

Today's Inventors and Their Inventions

Richard L. Gausewitz

ALSON PUBLISHING CO.
Long Beach, California

Printed in the United States of America.

Library of Congress Cataloging in Publication Data

Gausewitz, Richard L. (Richard Lord)
Patent pending.

Includes index.
1. Patent laws and legislation — United States. I. Title
KF3114.G38 1983 346.7304'86 82-9341
347.306486

ISBN: 0-916943-00-3

Production coordination: Del Mar Associates
Copy editor: Jackie Estrada
Book design: Louie Neiheisel
Cover design: John Odam
Composition: Boyer & Brass

DEDICATED TO

Ruth W. Jones
Mary-Ellen Gausewitz
Lynn Ellen Torrini

Preface

A prominent patent lawyer recently observed: "For decades the U.S.A., the world king of the technology mountain, sat upon a pedestal on top of the technology mountain. The U.S.A. has been knocked off the pedestal but not off the mountain."

The fact that America's technological position in the world has deteriorated has become common knowledge. The reasons for this decline are, however, less well known. This book explores one of the more important: a patent system that did not give inventors, or their financial backers, sufficient assurance that inventions would be protected against copying.

The patent system is, in a sense, a paradox since in a society in which monopolies are anathema, it authorizes the granting of a temporary "monopoly" for the manufacture of a new product. Unfortunately, since the 1930s a host of well-meaning persons, failing to understand how inventions are made and commercialized, have attacked the patent system. Writing a few years ago about America's bicentennial and the American patent system, a man with great experience in both business and government asked: "Why . . . in view of . . . very strong and unqualified endorsements of our patent system at the highest levels of government from the beginning of the Nation to this very day — why is it necessary constantly to repeat and publicly rededicate ourselves to continuing . . . that system as being so vital to our way of life?" The answer, he thought, was in the widespread *lack of knowledge* about its functioning.

Although hundreds of thousands of people are directly involved with or affected by what happens in the field of patents, not one in a hundred has an adequate understanding of what patents are all about. There is even less comprehension of the tests and other criteria used to determine whether or not a particular patent will be sustained in court. Persons about to undergo major surgery probably have a better realization of the procedures, risks, and "rewards" than have inventors and investors preparing to subject a patent to the rigors of a court battle.

It is in large part through court tests that the health of the patent system, and of the American inventor and entrepreneur, is measured. Surprisingly, the most famous and significant court cases in recent decades have involved people whose inventions were not complex or highly technological, but easily understood, mundane, and simple. The court decisions that resulted have, nevertheless, been subsequently applied to all types of patents, including the great majority that cover complex and technological inventions.

This book is written in the hope of shedding some light on the patent system, on the strong personalities that have conducted the battles within it, on the amazing concepts that have resulted, and on the unexpected renaissance that has recently occurred.

Richard Lord Gausewitz
Orange, California

Contents

1	The Obvious Girdle	1
2	The Incredible Tangle	23
3	The Fallout from A&P	43
4	One Up, Two Down in the Supreme Court	53
5	Damage in the Lower Courts	73
6	A Big Winner	89
7	The Road Toward Confusion	105
8	The Cow Barn Disaster	121
9	Why the Unequal Balance?	149
10	Vastly Better Patent Pending	185
	Appendix	219
	Source Notes	221
	Index	237

Patent Pending

ONE

The Obvious Girdle

Originality is rare:
most of us know a good thing
as soon as
the other fellow sees it.
EVAN ESAR

Being an originator does not make you an inventor. If that statement sounds paradoxical, imagine hearing it from a federal judge. Imagine, for example, how Olga—the highly successful fashion designer, entrepreneur, and inventor—must have felt when she found herself among the ranks of other inventors who had enjoyed the financial (and psychic) rewards of commercially successful inventions only to wake up one day and find their patents declared invalid. When this happened to Olga—when the court called her, in effect, "only" an originator—she could not believe it. She felt as if someone had proclaimed her a foot shorter than she actually was. She was stunned.

Olga was justified in being shocked by the court's decision. Not, however, because of the often-held belief that all "originators" deserve patents. They don't. Rather, Olga was caught up in a legal system that had evolved to a point where most patents did not afford the degree or kind of protection that they were meant to. The reasons for this evolution are subtle and complex. They can perhaps best be understood by examining a series of people (and their inventions) involved in courtroom misadventures.

Fashion-conscious women will recognize Olga's name immediately. She is Olga Erteszek, the famous designer of women's lingerie, whose handsome photograph graces the ads proclaiming that "Behind every Olga there really is an Olga." A regal woman of continental charm and courtliness, five feet seven inches tall, with blue eyes and auburn hair, she is always seen smiling enthusiastically next to her latest fashion creation. And, if you look closely at the capital letter "O" in Olga, you will see it broken by a small dove carrying a flower in its beak. This is the trademark that Olga and her husband, Jan, created twenty years ago to connote beauty and freedom—the political freedom they found when they arrived in California from war-torn Europe forty-one years ago with only a few dollars in their pockets.

The story of how the Erteszeks turned those few dollars into a company whose net sales for 1981 were a record $58,168,000 is, as the apparel industry trade press loves to remind its readers, in the best American "rags-to-riches" tradition.

Jan and Olga had grown up in Krakow, Poland's lovely medieval capital. Jan's father, assisted by his highly imaginative and brilliant wife, was a reasonably affluent businessman and apartment house owner. In 1932 Jan, nineteen years old, enrolled in the University of Krakow law school. That same year he received the first of several patents (for a commercially unsuccessful wooden fountain pen) and met Olga Bertran. She was sixteen, a student at the Emilia Platter School for Girls, a classical European *Gymnasium* that required eight years of Latin and four of Greek in addition to mathematics, geography, history, literature, and German.

Olga's mother was an expert corsetiere, who designed and sold corsets in her small shop on prestigious Szewska Street. There the teenage Olga earned makeup and candy money by fashioning boutonnieres and corsages out of silk.

From the time she was a young girl, Olga displayed a talent for visualizing. Sitting at the feet of the family's dressmaker, she made doll clothes that were copies of the clothing the woman sewed for her mother and two sisters. After graduating from the Platter School, Olga went to work full time in her mother's shop. Her speciality was designing postoperative girdles for the officers' wives in Krakow's military hospital. "I was quite artistic and very quick to learn," she recalls, "but I can't draw a straight line. I suppose that's the reason I went into curves. I can't paint, and I can't draw, but I can do anything with scissors. I can cut out and I can visualize things. I know, for example, what a garment is going to look like ahead of time, as I conceive it in my mind. I think I would have made a good architect because I have a great sense for lines as well as for colors and combinations."

This extraordinary ability to visualize is a central theme of invention, for, as the historian Eugene Ferguson has written, innumerable inventions, from the Newcomen steam engine and the mercury-vapor lamp to the snowmobile and the axe, owe their origins to nonscientific and nonverbal thinking and reasoning. "Pyramids, cathedrals and rockets exist not because of geometry, theory of structures, or thermodynamics, but because they were first a picture—literally a vision—in the minds of those who built them."

In 1938, after Olga had been working for her mother for a little less than a year, Jan received his Doctor of Laws degree and they became engaged. A job with a Polish government agency brought him to New York City. In August 1939, as war grew near, Jan hurried home to marry Olga. On September 1, Hitler struck, and before the month was out, Poland was divided between Germany and Russia. Fleeing eastward from the Germans in search of an American official to assist them, the young couple eventually arrived in Soviet-occupied Lvov. There they remained for a year, two nights of which Jan spent in a prison camp, because, he shrugs, "they didn't like me." Eventually, with the help of the American consul, Jan was able to leave. In December 1940, he headed, via the Trans Siberian railway and Vladivostok, toward his brother's home in California. It was only after he reached San Francisco that the Soviet authorities permitted Olga to apply for her visa. It was granted in time for her to leave the following March.

The Erteszeks were virtually penniless when they rented a small apartment in downtown Los Angeles. With Jan unable to practice law, Olga took a $16-a-week job with a small brassiere manufacturer. There she learned to operate a power sewing machine. She also learned that her boss disliked being told how the product might be improved. When her request for a $2 raise was rejected, she quit, convinced that she could do better on her own.

A number of people who have met Olga suspect that her obvious design talents are accompanied by a shrewd grasp of business. Nevertheless, she herself insists, "I was not meant to be a businesswoman; I still am not a businesswoman, and I'm thankful to the fates that my husband is a businessman, because I would have never been able to do it on my own. We made a good pair."

It was Olga's glimpse of a woman boarding a Los Angeles streetcar that set in motion the family partnership (later dubbed "sort of a two-way stretch" by a pun-happy *Los Angeles Times* reporter): "As the woman climbed those high steps she revealed her hose rolled up underneath her knees. To me this was disgusting for a well-dressed woman—it reminded me of the peasant women in Poland. I went home and told my husband I did not understand why they didn't wear garter belts."

Jan asked, "What are garter belts?"

Olga explained they were simply narrow girdles encircling the waist and holding up hose by means of elasticized tab garters.

Jan, always enthusiastic about Olga's ideas, suggested, "Why don't you make one?"

They calculated that the $10 remaining in their bank account would cover one month's rental for a sewing machine, a few yards of good fabric, and some dainty cloth rosebuds — the special Olga touch. "I didn't want the garment drab-looking — a woman needs something very feminine and very pretty," she explains.

The Erteszeks' apartment at the corner of Beverly Boulevard and Vermont Avenue consisted of a living room with a Murphy bed in the wall, a tiny kitchen, and a bath. The sewing machine was placed in the kitchen, and Olga set about turning out garter belts. After her first collection was ready for display, Jan was designated salesman. They chose the fashionable Bullock's Wilshire department store, only eight blocks away, as the first target. Jan, a dignified, slightly embarrassed European Doctor of Laws, stuffed a few garter belts into his jacket pocket and set out in the hope of finding a friendly buyer. The Bullock's buyer liked what she saw, but was amused when Jan naively quoted her a price for a single belt. She gave him a quick course in merchandising and sent him home with a $33 order for two dozen garter belts — to be delivered in only two weeks. Olga managed to meet the deadline, only to discover that American stores had a disturbing custom: instead of paying upon delivery, they paid on the tenth of the following month, and then took an 8 percent discount. Fortunately, a cash-on-delivery order for belts for a neighbor who had chanced to hear the whirring of Olga's sewing machine supplemented the Erteszeks' meager budget until the tenth. When payday arrived, they had enough money to buy fabric for a second order, this time for four dozen belts.

So the Olga Company originated — tentatively, fragilely,

one month before Pearl Harbor, its progress marked in dozens of garter belts. Soon there were enough orders to require the renting of a room in a nearby building. Jan had to learn to handle shears and cut fabric, as well as wrap and deliver packages. Business grew; helpers were added; the one room expanded into an entire floor. Inevitably, crises befell the fledgling enterprise. "Every day was a day when you wondered whether you were going to make it," says Jan. "It was obvious that we could not out-produce, out-sell, out-organize, out-administer, out-finance or out-price. The only thing we could do was out-think and out-imagine—out-innovate—because that didn't cost any money."

Garter belts proved very profitable, but as the war drew to a close Jan became uneasy over "the great dangers in being concentrated on anything exclusively in a rapidly changing world." Girdles, which were not radically different to manufacture, were a natural direction in which to expand. Even so, Jan remembers that he did a lot of soul searching. As a small entrepreneur, "If you make a wrong move you are out of business."

Against the advice of buyers for the big stores, the company initiated the new line. Using her own figure as a model, Olga designed her first girdle. It was an immediate success. Within a few months, Jan knew he had been right. Girdles became the firm's major product for the next quarter century.

Olga succeeded in the girdle business by pushing her inventive talents to the utmost. The traditional woman's girdle was basically an elastic sheath that encircled the torso below the waist. The idea was to "correct" the figure, principally by flattening the stomach. As Olga later explained in federal court, whereas other parts of the anatomy had been emphasized in various ways over the centuries, "there was never a time in history when stomach was fashionable." Essentially, the desired figure control was achieved by cleverly piling on layers of elastic and stiff fabric, pushed together and fastened with stays, bindings, and zippers. The stiffer the fabric, the better it flattened obstreperous rolls and bulges; so it is not surprising that the 1940s girdle, which began at the waist and came almost to the middle of the thigh, was a cumbersome and constricting garment, one that was particularly uncomfortable in warm weather.

Olga (and other designers) set out to devise improvements. Nylon and other new synthetic fibers helped enormously, but there is far more to making a girdle than simply substituting lighter weight fabric. Unlike a blouse or a slip, girdles, like bras, have to *do* something. They must support and shape flesh. Thus, like an engineer who designs a bridge, a girdle designer needs an understanding of stresses and strains, and of how forces act on one another.

One of Olga's earliest successes furnishes a good example of the ingenious engineering that corsetry requires. Olga was trying to solve the tricky problem of flattening the stomach without, at the same time, having the girdle cut painfully into the wearer's waist when she sits down. (If you can't visualize this, observe what happens when you try bending an aluminum can in half.) Her first step was to put a V-shaped "dip" in the front of the girdle so that it would not merely encircle the waist in a straight line, as conventional girdles did. Because the V had to retain its shape regardless of the position of the body, Olga devised an arrangement of five elastic panels that pulled in five different directions. The resulting French Secret girdle "dipped and skipped the waist," as Olga's ads put it.

In the years that followed, Olga created new designs and improved existing ones, along the way collecting thirty patents, many awards, and such laudatory newspaper appellations as "one of the greatest undercover agents in the world." Though girdles remained the company's mainstay until the early 1970's, it also branched out into bras and then into sleepwear and loungewear. In 1966 the trade paper *Women's Wear Daily* credited Olga's "intuitive fashion sense, coupled with skillful merchandising and marketing techniques" with building "an impressive business" in which sales were "booming" and volume "zooming." By 1967 the company was selling more than ninety styles of girdles, bras, and lingerie to some 2,300 retailers (including such top stores as Lord & Taylor, Bullock's, Bloomingdale's, and Saks Fifth Avenue) and was increasing its earnings at 20 percent annually. The Bank of America ran a full-page ad in the West Coast business press that showed an ethereal profile of Olga with teardrop earrings, facing the shoulder of her husband, who was looking confidently, with just a trace of a smile, past her at the camera. The photo was captioned: "We helped them produce some remarkable figures."

By the early 1960s, increasing numbers of women were switching to lighter, less formal and confining garments, including the panty girdle and its legless cousin, the panty brief. As far as Olga was concerned, however, "A brief was a brief was a brief. There was no deviation from it. It was a small panty girdle, legless . . . doing near nothing." Olga felt strongly that existing briefs, including her own, not only did nothing for the stomach, they were uncomfortably tight around the legs, particularly when their wearers sat down. In 1962, with her salesmen pleading that buyers for stores wanted a better brief, Olga once more put her philosophy, "If you're not just one step ahead of the others, then you are falling behind," into practice.

Her starting point, in trying to create a brief that would flatten the stomach and still be comfortable, was a baby's diaper. Setting out to turn the mental picture of folding a diaper from the baby's backside, through the crotch, and over its front like a napkin, into the resulting girdle trademarked Tumm-ee Breef, Olga followed her usual practice of locking herself alone in an office, away from interruptions, even by her husband. As various possibilities for simulating a diaper came to mind, she cut out the patterns herself, sewed them on the machine, and—"always my own guinea pig"—tried the briefs on.

Eventually she came up with the idea of sewing a large

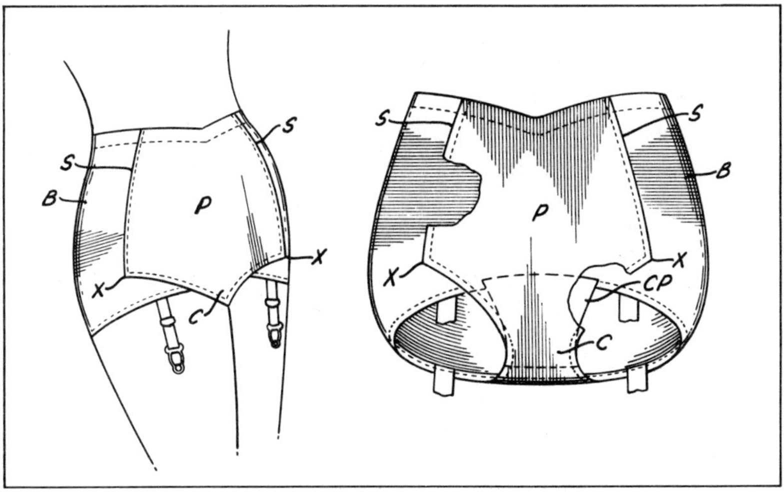

Olga's brief (the Tumm-ee Breef).

elastic panel (P in the drawings) to the *outside* of the elastic body B of the brief. This "tummy control" panel was attached to the body of the garment only at the top and the side seams and therefore "floated" quite independently of it. At its lower end the panel tapered inward, going between the legs to serve as a crotch piece C connecting to the lower back edge of the body B. By having the tummy panel "float," Olga was able to produce a cross-tensioning in the abdomen that flattened the stomach without exerting a painful pressure in the crotch. By extending the side seams S only down to points X, she obtained leg openings that adjusted to fit the wearer's legs. (Panel P was not attached to the front of body B below points X.)

Few inventions work perfectly the first time, and Olga's Tumm-ee Breef was no exception. When wearing the brief, Olga found that the loose lower edge of the body tended to roll up as she sat, leaving an unsightly wrinkle across the front of her dress. She solved this problem by connecting the outer crotch part C to the lower front edge of the body B by means of a separate inner crotch part CP made of lazy stretch nylon. In this way she provided an opposing force so that the garment body became "a part of the panty with the crotch pulling it down." She thus completed what a judge would later call "a unique and novel garment structure."

Commercial success for the Tumm-ee Breef was not only immediate (there were sales of 70,502 in 1963) but sustained. Over the next ten years, over 650,000 were sold to stores at an average price of nearly $4 each. When Olga's brief went on trial, an industry executive testified that such a life span indicates that a garment performs a fundamental or basic function rather than being merely a matter of transitory fashion.

Olga did her creating in a real "world"—that of inventing—which is but little understood by the average person. Perhaps even less understood (because they are not natural phenomena but the products of history and personalities) are two other fascinating elements to Olga's story: the patent system and the world of the courts.

At the Patent Office

The patent system was established by the United States Congress in order to fulfill a Constitutional purpose: "To promote the progress of . . . useful arts, by securing for limited times to . . . inventors the exclusive right to their . . . discoveries." Olga sought this exclusive right when in November 1962 she had her patent lawyer apply to the Patent Office* for a patent on her original tummy panel invention and, five months later, on the internal crotch piece that eliminated the roll-up problem.

Neophyte inventors often assume that obtaining a patent involves nothing more than "filling out a form." They quickly learn otherwise. A patent application is actually one of the most difficult legal documents to compose. Though inventors are free to prepare and file their own applications, the Patent Office warns that unless they have a great deal of experience

*The name is now Patent and Trademark Office, a part of the Department of Commerce. For brevity, the old name Patent Office is used.

they will be better off leaving the task to a patent lawyer or agent. (A patent lawyer is a lawyer with an engineering or science degree, or equivalent background, who specializes in patent work and has been registered with the Office after passing an examination. A patent agent is likewise technically qualified, has passed the test, and has been registered, but is not a lawyer.)

A patent application can be from three to one hundred or more pages long. It must include ink drawings prepared to precise specifications by an expert draftsman, a written description of the invention that is clear and complete enough to teach someone skilled in the particular "art" how to make it, and one or more "claims" that summarize all the essential features of the invention.

The claims are the crucial part of the application because they are the primary target of scrutiny at the Patent Office as well as in any courtroom in which the patent may subsequently end up. Writing claims requires the lawyer to choose language that is broad enough to provide adequate legal protection—against clever attempts to avoid the patent—and yet not so broad that the claims can be construed as describing earlier inventions. It is not too great an exaggeration to say that the writing of really sound claims requires the foresight of a prophet.

Before the patent application is written, however, a search is normally made to determine whether any prior invention exists that is identical to the one at hand. If it does, then there is no point in applying for a patent because the application will be refused for lack of novelty. The search can only be conducted using the astonishingly elaborate scheme by which the Patent Office has organized the 15 million American and foreign patents it stores in its eleven-story glass-and-steel Crystal Plaza complex in Arlington, Virginia.*

*The Patent Office has some 2800 employees and (in 1981) a $116 million annual budget.

Most lawyers assign this task to one of the several hundred patent lawyers and agents located in the Washington, D.C. area. These searchers are usually instructed to spend one to ten hours (at, for example, $50 per hour) looking for the closest "prior art," as patent lingo terms the works of earlier inventors.

A searcher begins by determining where in a system of 310 classes and about 108,000 subclasses, first established in 1919, the invention seems to fit. By consulting the six-inch-thick, fine-print *Manual of Classification*, a knowledgeable searcher can, for example, quickly track Olga's girdle to "Surgery, Bandaging" (the logic being that girdles change the shape of the body). There, amid some 65 subclasses, he will find the proper ones for each of her inventions under the entries: "523 . . . With individual buttock-receiving concavities," and "524 . . . With crotch or leg straps or leg-encircling parts." At this point he plunges into the long rows of six-foot-high stacks in the cavernous public search room, where almost 13 million copies of the more than 4 million American patents that have been issued are cross-referenced and available for study. He may also probe further by going upstairs and consulting with members of the Patent Office's examining corps. They will permit him to use their far more extensive files, which include over 10 million foreign patents and another 1 million pieces of patent literature (product catalogs, technical journal articles, and so on), all arranged according to subclasses. Because this horde is so vast, because it exists as individual pieces of paper rather than on electronically retrievable microfilm or computer tape, and because a patent may well be cross-referenced to other subclasses that, in turn, can lead to relevant patents in still other subclasses, a searcher embarks upon an essentially endless hunt that could cost thousands of dollars; he is constrained only by his budget.

On the basis of a searcher's discoveries, the inventor's lawyer advises his client whether to proceed with an application. If the decision is positive, the application is filed. A few months later (these days), the application reaches the patent examiner who specializes in that particular field. At the present time there are some 1,000 examiners, and they hug the perimeters of eight floors of two buildings in Crystal Plaza. In their midst the 24 million copies of United States and foreign patents and related documents, growing at the rate of at least half a million documents annually, are stacked in some 2 million shallow trays called "shoes."

Patent examiners are civil servants who have met certain requirements of technical education and competence. Examiners eventually become so familiar with their specialties that they can sometimes pick a related patent out of its shoe in just a few minutes. Essentially, an examiner makes the same kind of search as the patent lawyer, but, because he is more familiar with the prior art in his field, his search may be more complete. In examining Olga's girdle application, for instance, the examiner might have remembered a specific ad in an early Sears, Roebuck & Co. catalog, or an article in *Corset and Underwear Review* that had been filed away in a shoe along with prior United States and foreign patents.

At the conclusion of the examiner's initial search, there begins what is sometimes called "The Patent Office Minuet." It usually goes something like this:

1. The examiner rejects some or all of the claims in the application, basing the rejection on one or more of the prior patents discovered in the search. This rejection is not final, nor is it a mindless act. However, by long tradition (as well as Patent Office Rules), the rejection is well understood by both lawyer and examiner to present to the lawyer an opportunity to avoid the rejection.

2. The lawyer answers the examiner by arguing in writing—and often in person or by telephone—that a patent should be granted. Usually, he also amends at least some of the claims, changing their wording in manners calculated to "distinguish" them from the prior art.

3. The examiner then makes his decision to grant the patent, refuse to grant it, grant it in part, or refuse it in part. Most of the time the examiner doesn't reach this decision until the inventor's attorney has been heard at least once, and sometimes more than once. This is the final round of the minuet. With over 100,000 applications received each year, it is not surprising that the duration of the average examining procedure today is nearly twenty-three months.

On what legal grounds does the examiner make his decision? He looks at the numbered claims at the end of the patent application, each listing parts or "elements" of the invention. Had you, for example, applied for a patent on an early safety pin, you might have claimed: "A safety pin made of (a) a head, and (b) a return-bent hairpin-shaped spring wire, one end of which is connected to the head, and the other end of which is pointed and can be moved in and out of the head." If the examiner in the subsequent search had found a *single* earlier patent that showed *both* the head and the same kind of spring, working the same way, that would have ended the matter (because of "anticipation" by the single earlier patent). He would have refused to grant a patent for your combination of the (a) and (b) elements. In the world of patent law, it makes no difference that the person trying to get the patent was unaware of the earlier patent.

Usually, however, no *single* earlier patent shows *all* of the elements listed in any claim. In the case of the safety pin, the examiner might have found an ordinary straight pin in one patent, a metal hairpin in another, and a head in a third. He would then have had to decide (if today's laws had been in effect when the safety pin was invented) whether it would have been *obvious* to put the subjects of the earlier patents together in such a way as to make the safety pin. If he thought it would have been obvious, then he would have rejected the patent application for "obviousness."

Both of Olga's patent applications went through this ritual and were rejected for obviousness. After several months of back-and-forth arguments over earlier patents, including a "Girdle Panty Garter Belt" (Patent 2,763,008) invented by one William Rosenthal, the examiner finally granted both patents. The patents were then printed, enclosed in beautiful covers, sealed with large red seals, and tied with blue ribbons. They were issued on the same day, July 28, 1964, as Patents 3,142,300 and 3,142,301.

(Incidentally, had the examiner rejected one or both of Olga's applications, she could have appealed to the Patent Office Board of Appeals. Had she lost there, she could have brought a court action in the District of Columbia federal courts or—what was much more usual—appealed to the U.S. Court of Customs and Patent Appeals, a five-judge court whose wide experience in patent law made it the most patent-wise court in the United States.)

Three years after Olga received her two patents, Jan told an interviewer, his dismay possibly tinged with pride, "We are the most copied company in the business." An Olga Company executive recalls that most such disputes were settled readily, with the alleged infringer either discontinuing the product or taking a license, permitting him to copy the patented item, from the Olga Company. Consequently, none of Olga's previous patents had been infringed for substantial lengths of time. ("Infringing" usually means making, using, or selling an object or device containing *everything* listed in *any* single claim of a patent.)

But what happens when someone won't take a license under the patent (assuming that one is offered), and won't discontinue use of the patented invention? That, as Olga and Jan Erteszek were to learn, is when it is necessary to enter the world of the courts.

The World of the Courts

Vanity Fair Mills, a large, famous, and respected manufacturer of intimate apparel for women, had placed on the market a brief—number 40-6—not at all like Olga's. This brief turned out to be a disappointment. A number of buyers for stores dropped it for various reasons, one of which was lack of sufficient tummy control. According to Vanity Fair's vice president of sales, "it was obvious that 40-6 was not a good style and we wanted to establish a good brief style."

Vanity Fair consequently came up with a better brief, number 40-28, and by 1967 was selling it as part of its Taper Mate group of girdles. Its catalogs proclaimed "maximum control with minimum awareness of it," "power net ingeniously cut to prevent binding," "strategically self-reinforced at front then reinforced with a free power net panel which extends over tricot gusset." Between 1967 and 1969, when the brief was discontinued, more than 140,000 were sold and it was Vanity Fair's largest-selling brief.

The executives of Vanity Fair, after discussing Olga's Tumm-ee Breef, had instructed the design staff "to execute for us a garment which had the features that we were talking about, increased degree of span control at the waist and an improved leg opening." The designer admitted she had long been aware of the patented brief and that there was one stored in a carton near her working quarters at Vanity Fair. When asked whether the Taper Mate brief was original with her in regard to the crotch and panel construction, she replied, "I don't think there is anything new." (In fairness it should be noted that it is common practice in industry to obtain and study competitors' products, if for no other reason than to make certain that one's own product is better.)

Vanity Fair operated, at all times, with legal advice from one of the top patent law firms in the United States. Prior to marketing its new brief, the company sent it to the law firm, together with Olga's patents and her Tumm-ee Breef. Are we infringing, and are the patents valid, asked Vanity Fair? In reply, a prominent attorney made the following points:

1. Over the years the courts have held very few garment patents to be valid (legally good and effective).

2. Recent Supreme Court decisions in mechanical cases

have made it reasonable to expect that such patents will continue to be difficult to enforce in court.

3. The panty girdle field is old, leaving little room for unobvious developments.

In summary, the lawyer said that while Vanity Fair might be sued for infringement, it seemed likely—because of prior art patents—that Olga's patents would be held invalid.

Jan Erteszek's first reaction after becoming aware of the Vanity Fair brief was to write a personal letter to Vanity Fair's president advising him that Olga Company considered the brief to be an infringement of its patents. Erteszek noted he was calling the matter to the president's personal attention since ". . . I do realize that from time to time such situations occur inadvertently and that it is entirely possible that you are not aware of the existence of the patents to which I refer."

Vanity Fair's answer came eight weeks later—a lawsuit filed against Olga Company in the United States District Court, New York City, alleging that Olga's two patents were invalid. Such an action, which is possible *only* if there has been a threat such as Jan's letter, was standard operating procedure for somebody who did not intend to settle. By filing its own lawsuit—called a declaratory judgment action—Vanity Fair could pick the location of the trial. This was *very* important, since patent validity was often a matter of geography. Had the Vanity Fair president simply denied infringement, Jan Erteszek could have filed suit in a relatively favorable location.

It was never admitted by Vanity Fair that the Taper Mate brief was a copy of Olga's Tumm-ee Breef or that it was an infringement of Olga's patents. Nevertheless, the main thrust of its suit was to have Olga's patents declared invalid by the court because of "obviousness." The fact that a patent examiner has found an invention unobvious does not prevent a court from reaching the opposite conclusion. And if the patent is invalid, it cannot, of course, be infringed.

Although the suit was initiated in October of 1967, the

trial did not begin until January 29, 1973. Such a long delay is most unusual, and was caused by the backlog of cases in the crowded docket of the New York District Court. However, the intervening period did not lack for action, as it was taken up by the discovery phase of the litigation. During "discovery," lawyers for both sides attempt—in several ways—to uncover the facts. One such way is by depositions: the questioning of those with knowledge of the invention and its subsequent use. Deposition sessions are held in the lawyers' offices, where every word is taken down exactly as if the parties were in court.

"Discovery" is usually the most expensive phase of a patent lawsuit and can, in fact, almost bankrupt a small company. It cost the Olga Company thousands of dollars in attorney fees, plus many hours of management's time. Meanwhile, Vanity Fair was free to continue selling its brief, subject to payment of possible damages to Olga Company.

Eventually, however, the trial date was set, and what must have been one of the most interesting civil trials to occur in New York City in many a year opened in a drab courtroom high above a cold, gray, snow-pelted Foley Square. Only those directly concerned with the case sat in the dreary courtroom, with its dirt-streaked windows, grimy off-white walls, and government-issue walnut furniture. The judge was forty-two-year-old Thomas Poole Griesa, appointed to the bench by President Richard Nixon only six months earlier. Olga, quietly elegant in a business suit of blue wool shot through with white, insisted that, as a witness, she could explain the action of a girdle only on real flesh, not on a dummy. Consequently, a professional model was called in to wear the girdle over a black leotard. "A perfect twelve, beautifully proportioned and slender," according to one of the attorneys, the model went through various motions to demonstrate such features as comfort at the inner thigh, while the lawyers and witnesses attempted to maintain their dignity as they leaned forward to watch. Later, Judge Griesa said that he didn't fully understand a certain aspect of the operation of the Tumm-ee-Breef. To answer him, Olga made an unusual request of the judge, to which he graciously consented. She slipped the girdle onto his extended arms and pulled it in various directions so he could appreciate the effect it would have on a woman's legs. In the dialogue that ensued over pulls, panels, and tricot crotch pieces, she remarked: "We are trying to make your Honor an expert in the girdle business."

The same prominent New York patent lawyer who had advised Vanity Fair that it needn't worry about Olga's patents tried the case from start to finish, attempting to prove that earlier patents and advertisements had made it "obvious" for Olga to do what she did. Olga's lawyers in retort argued—as did Olga herself when she was a witness—that fundamental problems had been solved by the inventions and that earlier efforts to create stomach-controlling but comfortable briefs had been failures.

Of all the prior patents introduced in evidence by Vanity Fair, the patent for William Rosenthal's brief (issued to Maiden Form Brassiere Company and cited by the Patent

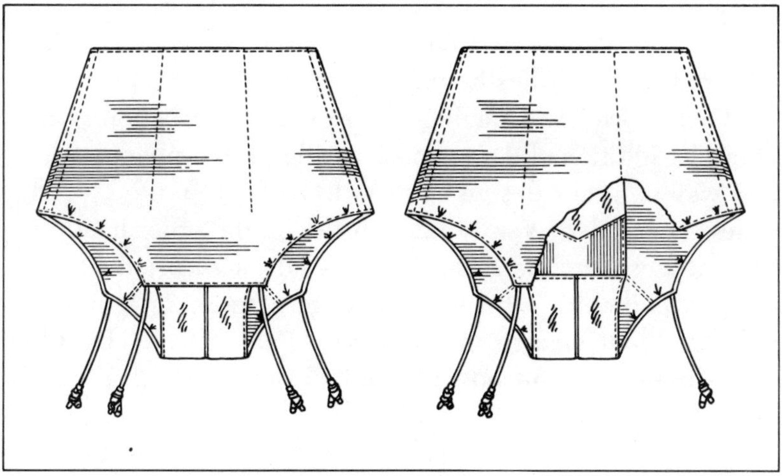

The Rosenthal garment, owned by Maiden Form.

Office in the initial rejections of Olga's applications) was the crucial one. The Rosenthal garment had a panel on the *inside* and differed from the Tumm-ee Breef in various other ways.

Olga was at a loss to understand why anybody would consider the Rosenthal garment relevant. In her testimony, she called it a "garter belt with a crotch inverted in it . . . [containing] a piece of dead fabric in the front which does not do anybody any good. To the contrary, it certainly couldn't be worn under any form-fitting garment because it would create a wrinkle." It could not be called a brief, she said, because "it cannot function without garters. . . . It is a regular garter belt, which means that the belt is dependent fully on the garters being attached to the hose. . . ." Regarding her own brief, Olga claimed, "garters were worn only in the case of a need for support for hose, because it was perfectly comfortable without them."

Judge Griesa sided with Olga. One year after the trial, he held Olga's patents to be valid and infringed. "I cannot agree that the Olga garment involved an obvious rearrangement of the elements of Rosenthal [Maiden Form girdle]," he ruled, and called the differences between the garments "far more fundamental than simply a change in position of a panel." He drew a sharp distinction between the types of leg openings, as well as "the degree and type of stomach control," which he declared to be "radically different in the two garments."

Olga's brief was "a highly original construction, which offered a solution of the stomach and leg and crotch problems in a way no other design appears to have attempted," Judge Griesa concluded. Vanity Fair's brief, on the other hand, was "essentially the same garment as the Olga . . . brief. . . . The unique, patented features of the Olga garment appeared virtually without change in the Vanity Fair model." Deciding that Vanity Fair's introduction of its brief in 1967 had caused Olga to lose $28,024 in royalties, he awarded Olga that amount plus interest for six years. (This award was distinctly low.)

Vanity Fair appealed the decision to the U.S. Court of Appeals for the Second Circuit. (The federal appellate court system is divided into geographical regions called "circuits.")* This court had only rarely upheld a garment patent, a record that Vanity Fair's attorney was undoubtedly aware of. After reading the transcript of the trial testimony, examining the exhibits, and considering the written as well as oral arguments by the lawyers, the three appeals judges issued their decision and opinion on February 4, 1975. (An appeal is decided not by the entire court, but by a three-judge panel.)

The opinion stated that Olga's "method of achieving vertical stretch maximization and independence of panel movement falls well within the capabilities of any person having ordinary skill in the art who gives serious thought to the root cause of the binding problems." This was a startling conclusion because, as the Court itself noted, "The troubling aspect of the case is that for well over twenty years no one prior to Mrs. Erteszek had been able to overcome the leg and crotch comfort problems which seemed endemic to this type of garment, thus casting some doubt on the obviousness of the solution." Nevertheless, the Court went on, "while the Erteszek patents solved problems that were 'basic' in the sense that they had persisted for a long time and thus seemed to be inherent in panty briefs, in the final analysis these patents represent style and comfort improvements rather than an innovative device for stomach control. Viewed in this light, the patented briefs have much in common with those designs which are 'new and pleasing enough to catch the trade,' but which do not reflect such exceptional talent beyond the skill of the ordinary designer as to merit a . . . patent . . . .

"This court therefore concludes that, while Mrs. Erteszek was an originator, she was not an inventor. Given the prior art and availability of assorted stretching materials, a skilled undergarment designer could by some concentrated thinking

*A map on page 219 shows all the circuits.

and experimentation arrive at the same results."

The "bottom line"—which a lawyer always looks at first (frantically) when a decision arrives in the mail—read: "The judgment of the district court is reversed." The patents were declared invalid, of no force or effect, and Vanity Fair was relieved of paying the Olga Company.

An *originator* but not an *inventor*? What the Court of Appeals was really saying was that Olga Erteszek did do something new but did not meet the standards required for a court to decide that a patent is valid. In other words, she was ruled to be the inventor of inventions undeserving of United States patents.

How could the Court of Appeals, admitting its own doubts in view of the factors in favor of unobviousness, have invalidated a patent granted by the Patent Office and upheld by the trial judge? Were patents to be sustained when the preponderance of evidence indicated an invention to be novel, useful, and unobvious? Or was the motto to be "When in doubt, invalidate"?

TWO

The Incredible Tangle

There being no existing basis of law,
the dictum of the judge
is one thing one day
and another thing another.

WILLIAM NEWTON
England, 1829

Persons whose knowledge of legal systems derives largely from Perry Mason, and whose familiarity with inventors comes from a mythologized Thomas Edison, see legal systems and inventors as models of logical thought and action. In court, brilliant lawyers and wise judges relentlessly pursue the facts and weigh them against clearly stated criteria until a just decision is reached. In workshops, inventors proceed carefully and methodically toward well perceived goals, steadfastly overcoming all obstacles until inevitably they triumph. Alas, life is far different. A major study (*The Sources of Invention*, by Jewkes, Sawers, and Stillerman) describes the reality of invention as "more often a series of stops and starts, of desperate frustrations and backtrackings, of logical steps intermixed with blind shots and of final success when it seems the most unlikely and the least hoped for." The same study characterizes the courts' search for rules to guide them in determining what constitutes a patentable invention as having produced "an almost incredible tangle of conflicting doctrines."

No wonder, then, that the real world of the inventor is light years away from the real world of patent law. Few judges understand or appreciate the process of invention. Few inventors understand the courts or know how to maneuver in a patent system rife with confusing legal theories, astonishing in both their nature and variety. However, the patent system itself has much underlying logic. It is only that some strange concepts have been grafted onto it in the United States (although seldom abroad). These concepts could easily be eliminated, thereby putting the whole system in an even better condition than the pristine one that existed before 1850, when all originators like Olga would have been considered inventors as well.

Pre-1850 America was a sprawling, brawling, pugnacious time of westward expansion, of nation building, and of inspired invention. The first half of that tumultuous century saw the flourishing of the inventive spirit that produced the electric telegraph, the mechanical reaper, the Colt six-shooter, the sewing machine, and so many more inventions that by 1876 a foreign visitor to the dazzling Centennial Celebration in Philadelphia exclaimed, "The American invents as the Italian paints and the Greek sculpted. It is genius." Genius it may well have been, not of the Einstein type but famed Yankee ingenuity. By protecting "genius" from the copying that could suffocate it, the patent system played no small part in the invention explosion.

A government's motivation for granting patents is simple. It is not, as some people think, to create private fortunes for inventors. Rather, it is to encourage individuals to risk years of their lives and scads of their money trying to make and market an invention that might benefit thousands or millions of people. There is a long tradition to this. When in 1594 the illustrious Galileo petitioned the Doge of Venice for a patent on an irrigation machine, he asserted the right on the basis of "it not being fit that this invention, which is my own, discovered by me with great labor and much expense, be made the common property of everybody." One hundred and twenty years before, the great city-state of Venice had passed what was probably the first general patent law, and Galileo stressed that the "favor" he was asking "is readily granted to any expert in every profession in similar cases."

From time to time various other European heads of state granted patents for inventions. In the second year of her reign (1559) Queen Elizabeth gave both a patent and a pension to an Italian named Acontio, who had invented certain furnaces and "wheel machines." Acontio's petition to the Queen said:

> [N]othing is more honest than that those who by searching have found out things useful to the public should have some of the fruits of their rights and labors, as meanwhile they abandon all other modes of gain, are at much expense in experiments and often sustain much loss.

The concept of patents came to the American colonies from England, and so clear did the benefits appear to the founders of the nation that there is no record of any debate in the Constitutional Convention of 1787 over the provision that granted to Congress the power to issue patents. As James Madison wrote in *The Federalist*, "The utility of this power will scarcely be questioned."

Congress passed the first patent act in 1790 and provided that patents be granted for up to fourteen years (they are now granted for seventeen years and, contrary to popular belief, cannot be renewed). In light of today's mega-government it is intriguing that under the 1790 act, the Secretary of State

(Thomas Jefferson), Secretary of War (Henry Knox), and Attorney General (Edmund Randolph) were personally responsible for examining all patent applications and granting patents for things they deemed "sufficiently useful and important." Not surprisingly, however, this practice only lasted for three years. After fifty-seven patents were issued (the first on July 31, 1790, to Samuel Hopkins of Philadelphia for "making of Pot ash and Pearl ash"), the cabinet members' burdens led to the passage of a new law that required no examinations at all.

Under this 1793 law, patents merely needed to be requested, much as copyrights are registered today for books, pictures, sculptures, and other literary and artistic properties. But this law pushed the pendulum too far, allowing patents to be granted for trivial and also ancient inventions. As a result, a new law was passed in 1836 that created a Patent Office and a Commissioner of Patents and that required the systematic review of patent applications by hired examiners. The examiners granted patents — if the commissioner deemed the invention "sufficiently useful and important" — for any "new and useful art, machine, manufacture, or composition of matter, or any new and useful improvement" thereon.

For the next fourteen years, whenever a dispute about an issued patent came into court, any originator who had complied with the 1836 law, and could demonstrate that no previous inventor had created the same thing, was assured of being ruled an inventor deserving of patent protection. The trouble was that there were still too many patents for trivial inventions. Something was needed to distinguish those inventions worthy of patent protection from those that were not.

The Supreme Court tried to remedy this situation in 1850, when it ruled (in effect) that in order to be validly patented an invention had to be more than new and useful — it also had to pass patentability tests *based on its merits as an invention.* If it did not pass these tests, the patent would be invalidated. Had the Court created only a *single, reasonable* test, all would have been well. But the Justices instead created *two* tests and thereby started to weave the incredible tangle.

The patent fought over in the landmark 1850 case could scarcely have been more innocuous. It dealt with a doorknob assembly in which the metal knob of the prior art was replaced by a knob constructed of ceramic. In voiding the patent, the Supreme Court (in an opinion written by Justice Samuel Nelson) said:

> For unless more ingenuity and skill . . . were required . . . than were possessed by an ordinary mechanic acquainted with the business, there was an absence of that degree of skill and ingenuity which constitute essential elements of every invention. In other words, the improvement is the work of the skilful mechanic, not that of the inventor.

This, a realistic test, should have been the sole one. But the Court also stated as its other test that a substitution of materials was of itself "never" patentable.

As the first Supreme Court case to require tests of patentability, the doorknob case eventually assumed a monumental importance that has continued to this day. How can a single court decision exert an influence that is felt over thirteen decades? Simply by the power of "precedent." In arguing cases, lawyers cite tests, key words, and phrases of the Supreme Court (and other courts). Indeed, as the foremost patent judge in the United States, Judge Giles S. Rich, formerly of the Court of Customs and Patent Appeals, observed, lawyers "seize on anything that is said in an opinion, especially one by the Supreme Court, that will help [their] case, blowing it up out of all proportion." In this way, and by their own research, the judges become aware of the language of earlier opinions (precedents) and may quote them in their own opinions. Thus, certain particularly famous tests and words are handed down to—and greatly influence—posterity.

Precedents by the Supreme Court are, of course, vastly

more important than those by the lower courts, but the Supreme Court rarely decides patent cases. In its unique role as the balance wheel for our tripartite form of government, the Court does not have the time to hear many of the numerous cases disappointed patent litigants ask it to consider. Only seven test-of-patentability cases have been decided by the Supreme Court after the year 1950. But, coming on top of certain earlier cases, these seven have set the direction for the lower courts, thereby affecting all patent litigants.

When a Supreme Court Justice is informed of a precedent, such as (for example) the doorknob case, he has a choice. He can follow the precedent and do much the same in the case before him, or he can ignore the precedent and generate new criteria (sometimes even a whole new patentability test) out of his own ingenuity. In the decades following 1850 Supreme Court Justices (and some other judges) found the temptation to create new tests too great to resist. Judge Rich once commented that the necessity to define patentable "invention" became "the plaything of the judges who, as they became initiated into its mysteries, delighted to devise and expound their own ideas of what it meant, some very lovely prose resulting."

The Supreme Court's opinions, subsequently applied as precedents in lower courts, evolved along two principal lines: (1) the articulation of certain "fixed rules" as to what was and was not a patentable invention; and (2) rationalization couched in catchwords to justify judgments that were actually products of a liberal admixture of "gut reactions" to the evidence.

Fixed Rules

Fixed rules originated with one of the tests of the doorknob case, where the Court said, "The improvement consists in the superiority of the material . . . over that previously employed in making the knob. But this, *of itself*, can *never* be the subject of a patent." (italics mine)

With this, the supposedly fixed, inflexible rule was created that a mere substitution of materials is not a patentable invention. The trouble is that some significant inventions can involve a substitution of materials. Some twenty-six years after the doorknob case, the Supreme Court backed off by declaring that the first good dental plate was patentable. A key to this invention was the use of hard rubber in place of the various materials that had previously been employed. In sustaining the patent, the Court discussed the doorknob case but said that it did "not decide that no use of one material in lieu of another . . . can, in any case . . . be the subject of a patent."

A far more famous substitution-of-materials invention was Edison's replacement of a platinum wire by a thin carbon filament, thus creating the first commercial incandescent light bulb. On October 4, 1892, the Circuit Court of Appeals for the Second Circuit held valid this claim of Edison's 1880 patent: "The combination of carbon filaments with a receiver made entirely of glass, and conductors passing through the glass, and from which receiver the air is exhausted, for the purposes set forth." The Court declared itself "satisfied that there was [patentable] invention in the substitution of the carbon of the patent in suit for the platinum of the [earlier] patent. . . ."

Other supposedly inflexible, fixed rules that the Supreme Court generated over the years following 1850 underwent similar treatment. Such rules as "mere change in degree is not patentable," and so on, all required "exceptions" as major inventions came along that violated the rules. Attempts to follow supposedly fixed rules continued for about a century after the doorknob case before going out of vogue. The great Judge Learned Hand once pointed out the difficulty:

> How is it possible to say a priori what combination of elements needs an original twist of the mind, and what is within the compass of the ordinary clod? . . . Our

unknown ancestor, who first substituted iron for bronze in the head of an ax, was the bright exemplar of all inventors to come.

Catchwords

The reliance on catchwords (and phrases formed from them) to define patentability originated with a case in 1870. In fact, the language seldom had realistic meaning in relation to the case being decided but became the most important part of the incredible tangle of doctrines regarding whether or not there was patentable "invention." Judge Rich observed: "How [a] decision was reached was rarely revealed. Everyone realized it was subjective."

The catchwords comprise a few words and expressions that evolved during the trials and appeals of a number of cases during the span of about a century. Closely related to one another, these catchwords were introduced in the pronouncements of famous Supreme Court opinions and were first heard in an 1870s patent battle between competitors in the Albany, New York, stove industry.

The stove being fought over was one of a type known as a "base burner." (The drawing shows a stove of this general type.) Base burners had come into existence a half century earlier when the Lehigh Coal Mine Company had asked one of the area's most prominent stove designers, the Reverend Eliphalet Nott, head of Schenectady's Union College, to devise a stove that would burn coal rather than wood. The Reverend Nott subsequently patented the "Saracenic Grate," an ornately decorated affair in which coal was poured in at the top and directed to the base for burning.

Unfortunately, Reverend Nott's Saracenic Grate turned out to have the disconcerting habit of exploding. It was not until 1855 that D. G. Littlefield of Albany made what seems to have been the first commercially successful base burner. Between 1861, when his Littlefield Stove Co. brought out an improved version called the "Morning Glory," and 1871,

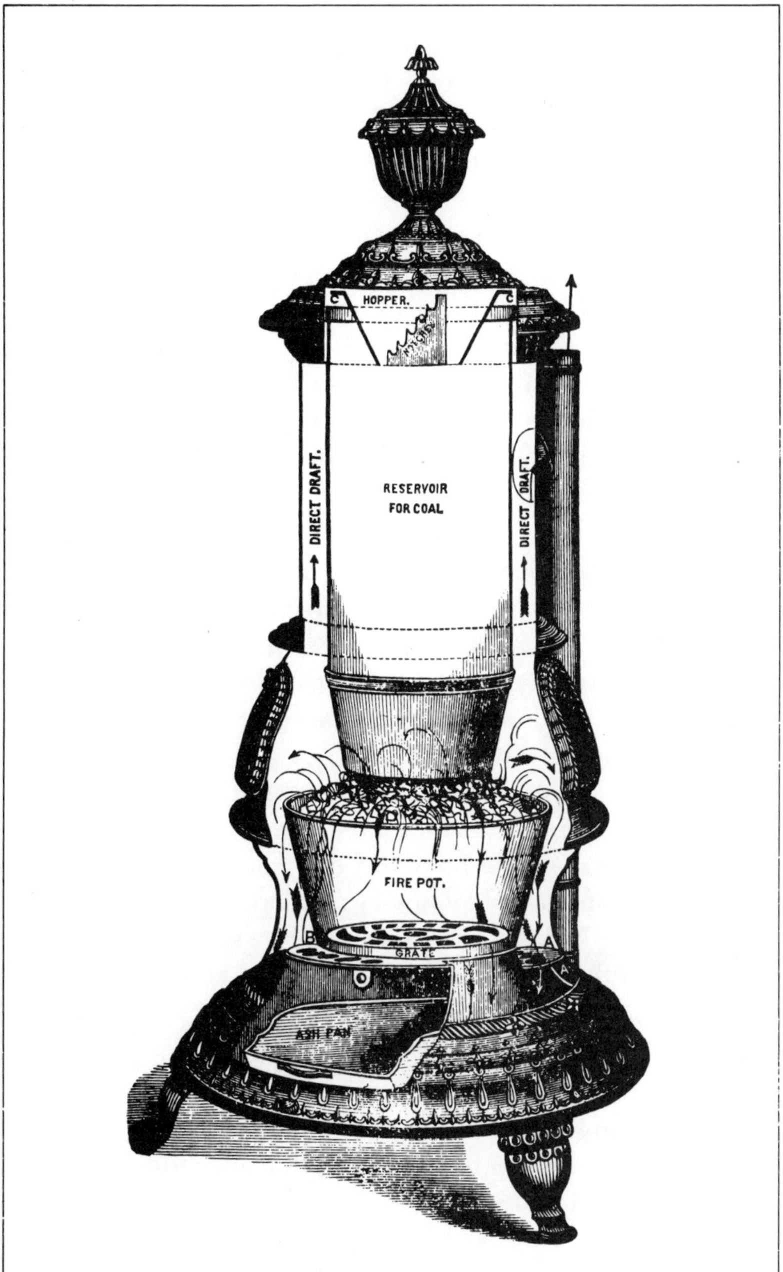

A base burner stove.

when he felt compelled to publish a thirty-five-page pamphlet giving "a full and correct history" of the base-burning stove, Littlefield sold close to 100,000 Morning Glory stoves.

Littlefield was not included in any court case described in this book. His pamphlet, however, is interesting for the background it gives. In it he inveighed against "various grades of infringers" with whom he had to cope. Some, he said, falsely represented themselves as his agents, while others "have taken my property to fix up the exterior portions of their stoves, with as much coolness as though they expected the approbation of the public for so doing." Still others he described as "honorable manufacturers, who seek in good faith to use only their own inventions, together with that which has become the property of the public. With such manufacturers there is never any serious trouble, for when, by accident, they infringe, they are ready to change and avoid interference, or pay for what they have inadvertently taken." From Littlefield's account, one senses the highly competitive, very likely cutthroat rivalry among Albany's fifteen stove manufacturers, who employed a total of nearly 2,000 men (at an average wage of $12 per week).

One of these enterprises in the dawn of the golden age of American industry was that of Jasper Van Wormer and Michael McGarvey. As boys the two had been apprentices together, and in 1846 they opened a shop with no capital but "indomitable pluck and an honest purpose." Their business prospered to the degree that in 1867 they were able to build their own four-story brick building and foundry. Within four years they were employing 80 men and generating sales of $250,000 per year. Then, however, Hailes and Treadwell, another Albany company that purveyed stoves, filed suit against Van Wormer and McGarvey for allegedly infringing patents obtained in 1863.

On the third Tuesday in January 1870, Judge Woodruff of the Circuit Court for the Northern District of New York dismissed the complaint. A "mere *aggregation* of devices," he wrote, was different from "a patentable *combination* of old devices which produces a new result or an old result in a better or more economical manner." (italics mine) Judge Woodruff's verbal creativity in making this curious distinction has haunted inventors ever since.

Hailes and Treadwell appealed directly to the U.S. Supreme Court, which was possible in those days. After pondering the written as well as oral arguments, Justice William Strong wrote an opinion agreeing with the lower court. Despite the fact that the case was dismissed primarily on the grounds of no infringement rather than patent invalidity, Justice Strong concurred that "a new combination" was patentable only if it produced "new and useful results [which] . . . must be a product of the *combination*, and not a mere *aggregate* of several results each the complete product of one of the combined elements. . . . Merely bringing old devices into juxtaposition, and there allowing each to work out its own effect without the production of something novel, is not invention. No one by bringing together several old devices without producing a new and useful result [which is] the joint product of the elements of the combination and something more than an *aggregate* of old results, can acquire a right to prevent others from using the same devices. . . ." (italics mine)

What Justice Strong probably meant by this outburst of verbiage was that he thought the different parts of the patented coal stove *did not cooperate or interact* with each other. But he did not say so in plain English, using instead the words "aggregate" and "combination." To Edmund J. Kitch, who wrote about the case in 1967 when an assistant professor of law at the University of Chicago, Justice Strong's aggregation-combination doctrine "managed to combine the appearance of complex significance with the substance, if not the grace, of a nonsense rhyme."

That aggregation and combination were being used to distinguish between absence and presence, respectively, of cooperation was made clear by a Supreme Court decision that occurred in 1875. The case was Reckendorfer vs. Faber (in all probability the Eberhard Faber who established a New York City branch of the European "House of Faber" in 1851 and gave us A. W. Faber pencils). Better known among patent attorneys than Hailes' coal stove, the case involved two patents, one of which was titled "Combination of Lead-Pencil and Eraser." The rubber eraser was mounted in one end of a wooden pencil, and both it and the lead at the opposite end could be sharpened.

Justice Ward Hunt declared the patent invalid. In an opinion supported by six of the nine members of the Supreme Court, he wrote, "This combination consists only in the application of a piece of rubber to one end of the same piece of wood which makes a lead-pencil. It is as if a patent should be granted for . . . a stick twelve inches long, on one end of which is an ordinary hammer, and on the other end is a screw-driver or a tack-drawer. . . ." After citing several cases, including that of Hailes' coal stove, Justice Hunt wrote, "the combination, to be patentable, must produce a different force or effect . . . from that given by their separate parts. There must be a new result produced by their union: if not so, it is only an aggregation of separate elements."

Despite its reference to "combination" and "aggregation," Justice Hunt's opinion made clear that what was necessary for patentability was that the parts *cooperate* or mutually interact

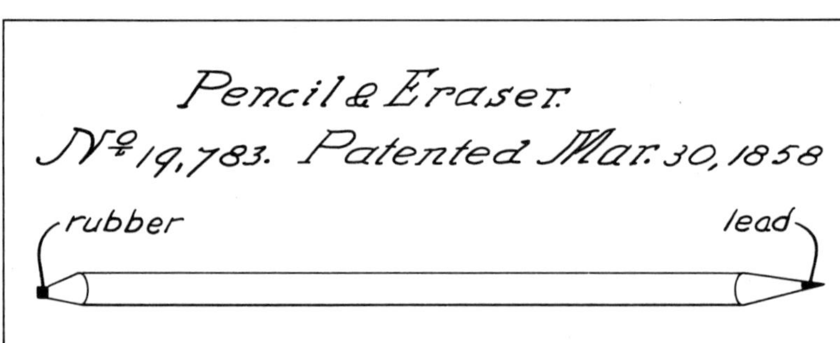

to produce a result. This is evident from his illustrations of *patentable* inventions, which included:

> the frame in a saw-mill which advances the log regularly to meet the saw, and the saw which saws the log; the two co-operate and are simultaneous in their joint action of sawing through the whole log.

After the illustrations, Justice Hunt wrote, "In these and numerous like cases, the parts co-operate in producing the final effect, sometimes simultaneously, sometimes successively. The result comes from the combined effect of the several parts, not simply from the separate action of each, and is, therefore, patentable."

Though the catchwords "aggregation" and "combination" were adopted by courts in patent cases throughout the United States, they soon became dissociated from the facts of the Supreme Court case that generated them. As Judge Rich lamented many years later (when writing about the evolution following the doorknob case): "As is usual with a doctrine derived from a court opinion, the doctrine persists and evolves while the facts of the case are forgotten." Thus, only six years after the pencil-eraser case, the Supreme Court used aggregation-combination to wipe out a patent for an invention in which the parts did, beyond question, cooperate.

In the decades that followed, right up until 1950, if a judge considered an assembly of old elements patentable he often labeled it a "combination." If not, he called it a mere "aggregation." The presence or absence of cooperation between the parts was *not* a factor in the overwhelming majority of these decisions. The words had become mere catchwords or labels used to rationalize decisions reached, not on the basis of the totality of the evidence, but largely on the judge's subjective feelings concerning whether or not the invention deserved a patent.

Then in 1950 came what was by far the most famous case

to be decided since that of the doorknob. Known as the "A&P case," after the supermarket chain a patent owner sued, it also involved an everyday object: a checkout counter.

In early supermarkets, which began to supplant old-fashioned grocery stores during the Depression, the checkout counter was usually shaped like an L, and the cashier and cash register were located at the elbow of the L. Each customer took her or his groceries out of the cart or basket and placed them on the counter to be checked by the cashier. The next person in line had to wait before placing her groceries on the counter. The delays that resulted produced a widespread demand for some system to speed up the process.

It fell to Ellis Turnham, an employee in a small Texas supermarket chain, to conceive of extending the counter and constructing a three-sided pusher, or tray, that used the top of the counter for a bottom and was prevented from slipping off by guide rails on each side. A handle permitted the cashier to pull the tray with the groceries forward. After the tray was slid back, leaving the groceries in front of the cashier, the next customer could place her selections in the tray.

Turnham installed his "merchandise handler" in the supermarket where he worked, and received a patent for it along with much publicity. When A&P began using the apparatus shown in the drawing, Turnham's employers sued, claiming infringement of his patent (which had been assigned to them).

The allegedly infringing construction had the counter extension at A, guide rails at D, the open-bottom tray at C, and the handle for moving the tray at E. A&P made no pretense of having invented this device. It simply charged that similar devices had been used before and that anyone "skilled in the art" could have perceived the solution once the need for it became apparent. The trial judge disagreed and sustained the patent. His decision was affirmed by a court of appeals.

The Supreme Court's invalidation of the patent was deliv-

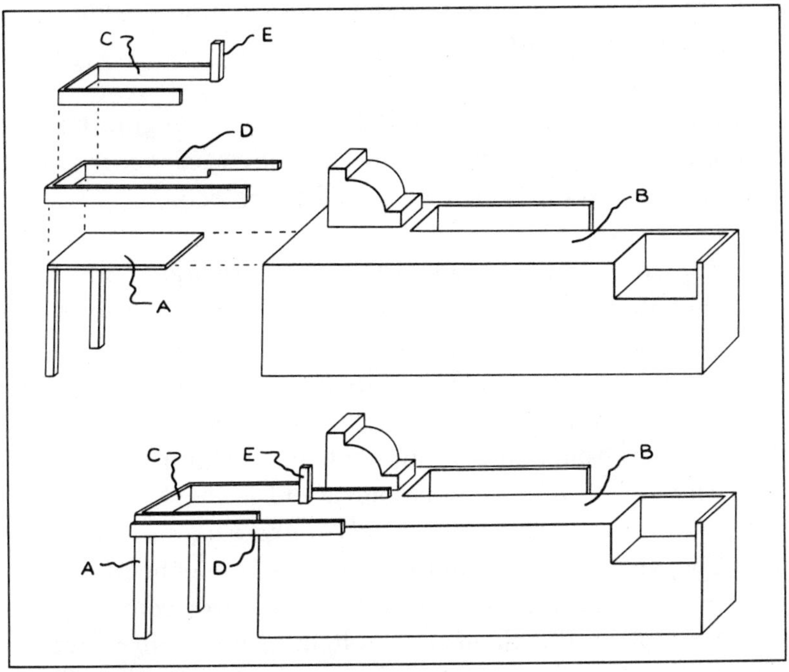

A&P's checkout counter.

ered on December 4, 1950 by Justice Robert Jackson in a brief but eloquent opinion that crammed more quotable antipatent language into a small space than has been done before or since. First, he displayed his suspicion of combinations of old elements by asking, "What indicia of invention should the courts seek in a case where nothing tangible is new, and invention, if it exists at all, is *only* in bringing old elements together?" (italics mine) Then he went on to attack the nineteenth-century concept of aggregation:

> The key to patentability of a mechanical device that brings old factors into cooperation is presence or lack of invention. In course of time the profession came to employ the term "combination" to imply its presence and the term "aggregation" to signify its absence [but] their

> employment as tests to determine invention results in *nothing but confusion.* (italics mine)

Having thus demolished the aggregation-combination catchwords, he promptly—in this and other passages—added to the confusion to which he so rightly objected:

> The conjunction or concert of known elements must contribute something; *only when the whole in some way exceeds the sum of its parts* is the accumulation of old devices patentable. Elements may, of course, especially in chemistry or electronics, take on some *new quality or function* from being brought into concert, but this is not a usual result of uniting elements old in mechanics. This case is wanting in any *unusual or surprising consequences* from the unification of the elements here concerned, and there is nothing to indicate that the lower courts scrutinized the claims in the light of this rather severe test. (italics mine)

The last quoted passage is incredibly famous and important, but of no benefit to patent owners. Along with certain other parts of Justice Jackson's A&P opinion, it has confused judges and lawyers and has hurt inventors. It is necessary to understand why this is so.

There was, of course, no intent to generate confusion and harm. Justice Jackson was striving to create a good test. And he was an intelligent, respected Justice and no habitual detractor of patents. In another case, he made a ringing denunciation of the Supreme Court's invalidation of a patent for a simple mechanical invention.

The explanation of the confusion and harm resides in (1) the nature of Turnham's checkout counter invention, and (2) the way lawyers and judges operate when applying precedents.

Justice Jackson had a low regard for Turnham's invention, probably agreeing with this assessment by one lower-court judge who dissented when the patent was sustained: "Quite apparently, Turnham derived the idea underlying his patent . . . by utilizing a commonly known and long existent art, that of racking pool balls upon a pool table." With an invention before him that he believed (whether or not correctly) to be trivial, it is natural for Justice Jackson to have stated things that he would have left unsaid had the invention been one he respected.

But when the precedent Justice Jackson set with this quoted passage was applied in subsequent years, the subject matter of the A&P case was disregarded. All that was used was the passage itself. Worse yet, it became common for lawyers and judges to dissect the quoted passage, pick only certain phrases, and apply them as "tests" of patent validity! The majority of these tests were really catchwords to seemingly justify decisions reached subjectively.

The "tests" resulting from the dissection are italicized in the quoted passage. Starting at the top, there is the most famous (or infamous, to those striving to sustain patents) statement: "only when the whole . . . exceeds the sum of its parts is the accumulation of old devices patentable." Since nothing exists outside of chemistry (or biology) in which the whole actually exceeds the sum of its parts, literal application of this test would mean that no mechanical or electrical invention could ever be patented.* For example, when the Wright brothers took off at Kitty Hawk, someone of philosophic bent could have exclaimed about their airplane that "the whole exceeds the sum of its parts." But it didn't *literally;* all they had was fabric, engine, frame, propeller, and so forth. Of course, the parts of the airplane cooperated with each other,

*Many argue that even chemical inventions cannot pass this or certain other of Justice Jackson's tests.

but so did the parts of the checkout counter.* And, under Justice Jackson's new test, cooperation between the parts—such as would have satisfied Justice Hunt in the 1875 pencil-eraser case—no longer sufficed. Future inventions were to be judged on the basis of whether the whole in some way exceeded the sum of its parts, a requirement infinitely more severe than that for cooperation.

Coming to the next sentence in the passage ("Elements may, of course, especially in chemistry or electronics, take on some new quality or function from being brought into concert, but this is not a usual result of uniting elements old in mechanics"), it sounds as though Justice Jackson wanted the parts to be somehow different in the combination than out of it. However, except for cooperation (which did not suffice) this cannot *literally* occur in nonchemical inventions. Only in chemistry, as when hydrogen and oxygen combine to make water, or sodium and chlorine to make salt, can the elements be different in the combination than out of it. Even in the Wrights' airplane, the propeller was—when cooperation is not taken into account—the same up in the air as when lying on some shelf. That Justice Jackson recognized this, in part, is evident from his statement, "this is not a usual result of uniting elements old in mechanics."

Finally, there is the last sentence in the famous passage: "This case is wanting in any unusual or surprising consequences from the unification of the elements here concerned, and there is nothing to indicate that the lower courts scrutinized the claims in the light of this rather severe test."

*Justice Jackson recognized that cooperation was present, for he wrote: "This counter does what a store counter always has done—it supports merchandise at a convenient height while the customer makes his purchases and the merchant his sales. The three-sided rack will draw or push goods put within it from one place to another—just what any such a rack would do on any smooth surface—and the guide rails keep it from falling or sliding off from the counter, as guide rails have ever done." The guide rails guided the rack much as the frame of Justice Hunt's sawmill advanced the log into the saw in the example from the pencil-eraser case.

"Unusual or surprising consequences" is still used as a test in some parts of the United States, but it is a poor test of patentability. It is a common occurrence for someone to look at an invention—especially a mechanical one—and instantly recognize the consequences, wondering, "Why didn't somebody think of that sooner?" (The consequences of the safety pin, for example, were neither surprising nor unusual. Anyone, *after* seeing the device, would know that the head would keep the pin from stabbing the baby.) Worse, the test has little to do with how difficult it is for the inventor to get the idea in the first place, then to select the parts from the mass of prior art, and finally to assemble them into something that fills a need.

Justice Jackson termed "unusual or surprising consequences" a "rather severe test." The word "severe" has been taken out of context and applied to other tests in the quoted passage by lawyers, judges, and commentators. Actually, these other tests are not only "rather severe," but impossible to pass—at least by nonchemical inventions—*if* realistic criteria are employed. Of course, Justice Jackson had no intention of imposing impossible tests; he was, instead, stating in extravagant prose some requirements that were to be difficult but not unattainable.

The basic trouble is that each of these new "tests," these additions to the incredible tangle, is—to this day—argued stridently by defense lawyers to judges as though they are *realistic* and should be applied *literally*. Yet, with the exception of "unusual or surprising consequences," the tests cannot literally be passed by any nonchemical invention. Thus, when judges have been influenced by such arguments, as has occurred too often, they have invalidated patents for many excellent inventions. One result has been discrimination against mechanical inventions, especially those that look "simple" to a judge.

In formulating his severe criteria, Justice Jackson evidently intended to define more clearly the line between what may be termed "worthy patents" and "shadow patents," terminology derived from an 1882 Supreme Court opinion that stated:

> It was never the object . . . to grant a monopoly for every trifling device, every shadow of a shade of an idea, which would naturally and spontaneously occur to any skilled mechanic or operator in the ordinary progress of manufactures. Such an indiscriminate creation of exclusive privileges tends rather to obstruct than to stimulate invention.

Justice Jackson, however, by predicating tests impossible for the great majority of (at least) mechanical inventions to meet, threatened to place a whole class of inventions in the category of "shadow patents" and merely compounded the problems stemming from the aggregation-combination test. Such "tests" as whole-must-exceed-sum-of-parts were more inclusive catchwords than aggregation-combination, and permitted judges even greater latitude in invalidating patents toward which they were unfavorably disposed.

The A&P tests, therefore, frustrated another portion of the 1882 Supreme Court opinion, which stated that inventors "who make some substantial discovery or invention, which adds to our knowledge and makes a step in advance in the useful arts . . . are worthy of all favor."

It was the A&P case that stimulated actions causing Congress to take the first of two steps necessary for reasonable, minimally subjective, nondiscriminatory separation of worthy patents from shadow ones.

THREE

The Fallout from A&P

It is idle to say that combinations of old elements cannot be inventions; substantially every invention is for such a "combination": that is to say, it consists of former elements in a new assemblage.

JUDGE LEARNED HAND

On December 5, 1950, Judge Giles S. Rich — then a New York patent lawyer — was reading from *The New York Times* to a group of patent experts at a meeting in Washington, D.C. He had their undivided attention because what he read was the account of the A&P case, decided the previous day.

Judge Rich and the other experts were members of a committee that was revising a draft of an entire new patent law, later passed by Congress as the Patent Act of 1952. The draft had been written by P. J. Federico, a distinguished examiner-in-chief, of the Patent Office. Federico had included an "obviousness" section, but the committee members disagreed as to whether that section should be retained or dropped. Said Judge Rich:

> In December 1950 the bar was far from unanimous in thinking that the statute should deal with the requirement for "invention," not even the members of the Committee agreed. There are always those who prefer the *status quo* with which they have learned to live, no matter how ridiculous it may be. Now, it is very significant that what persuaded the . . . Committee to replace the case law with a statutory provision was the Supreme Court's . . . "A&P" Case on the checkout counter.

What jolted the committee about the A&P case was not the result (the decision). And, it was not only such phrases as whole-must-exceed-sum-of-parts. It was, also, the express requirement that a more "exacting" test of patentability be used when a combination is made up of old components than when it is not.

At the end of his opinion Justice Jackson had written: "a standard of invention appears to have been used [by the lower courts, which upheld the patent] that is less exacting than that required where a combination is made up entirely of old components. It is on this ground that the judgment below is reversed." Another and often-quoted statement from the opinion is: "Courts should scrutinize combination patent claims with a care proportioned to the difficulty and improbability of finding invention in an assembly of old elements."

In the nineteenth century, however, the Supreme Court had taken the diametrically opposite view: "New elements . . . are not required, and if such a defense were allowed, not one patent of the kind in a thousand of modern date could be held valid." Thus, to impose a higher test (standard) of patentability for combinations of old components is to require a higher test for almost everything. Judge Learned Hand once observed, "It is idle to say that combinations of old elements cannot be inventions; substantially every invention is for such a 'combination': That is to say, it consists of former elements in a new assemblage."

Justice Jackson's opinion, therefore, was one more marker

in a century-long Supreme Court pendulum swing away from sympathy toward patents. Compare, for example, the Court's statement that "liberality, rather than strictness, should prevail where the fate of the patent is involved" in sustaining Charles Goodyear's patent for vulcanizing rubber in 1869, with Justice William O. Douglas' declaration invalidating the patent for the self-releasing cordless automobile cigarette lighter in 1941: "Strict application of [the skilled mechanic] test is necessary lest in the constant demand for new appliances the heavy hand of tribute be laid on each slight technological advance in an art."

Justice Douglas furthermore went off on a novel tangent, entirely at odds with the history of invention, by asserting that "the new device, however useful it may be, must reveal the flash of creative genius, not merely the skill of the calling. If it fails, it has not established its right to a private grant on the public domain." Because even Goodyear's vulcanization of rubber, Alexander Graham Bell's telephone, Edison's phonograph and incandescent light, and numerous other inventions were essentially the products of dogged persistence mixed with flashes of luck as well as creative genius, Justice Douglas' test—if applied literally—would have come close to eviscerating the patent system. After enjoying a brief vogue, the "flash of genius" test faded into the limbo that it deserved.

The mid-twentieth-century Supreme Court's lack of sympathy toward patents reached its dramatic height in 1943 when Chief Justice Harlan F. Stone delivered the five-to-three decision invalidating the broad claims of Guglielmo Marconi's 1904 radio patent. Justice Felix Frankfurter dissented acidly: "The inescapable fact is that Marconi in his basic patent hit upon something that had eluded the best brains of the time working on the problem of wireless communication. . . . The fact is that [earlier inventors] did not have the 'flash' (a current term in patent opinions happily not used in this decision) that begot the idea in Marconi which he gave to the world."

Justice Wiley Rutledge joined Justice Frankfurter, noting:

". . . as is often true with great inventions, the simplest and therefore generally the best solution is not obvious at the time, although it becomes so immediately it is seen and stated. Looking back now at Edison's light bulb one might think it absurd that that highly useful and beneficial idea had not been worked out long before. . . . The same thing applies to Marconi. Though what he did was simple, it was brilliant, and it brought big results."

It was for the two purposes of (1) moderating the extreme strictness toward patents exhibited by the mid-century Supreme Court and (2) eliminating the incredible tangle of patentability tests, that Judge Rich's drafting committee approved Federico's "obviousness" section in revised form. It became Section 103 of the Patent Act of 1952:

> A patent may not be obtained . . . if the differences between the subject matter sought to be patented and the prior art are such that the subject matter as a whole would have been obvious at the time the invention was made to a person having ordinary skill in the art to which said subject matter pertains. . . .

This section is *the* most important in the whole patent law. It is based roughly on the "skilled mechanic" test of the 1850 doorknob opinion and also on a few subsequent Supreme Court opinions that used the word "obvious" instead of relying on fixed rules or catchwords.

Writing the new test of "obviousness" into the law was one matter; implementing it was another. Judges, like other human beings, are the sum of their experiences. Justice Douglas wrote in his autobiography that he had "thought of the law in the terms of Moses—principles chiseled in granite." He thus found it "shattering" when Chief Justice Charles Evans Hughes told him he must remember, "At the constitutional level where we work, ninety percent of any decision is emotional. The rational part of us supplies the reasons for supporting our predilections." Douglas eventually recognized the truth of this statement. "Judges are, after all, not creative figures; they represent ideological schools of thought that are highly competitive."

Such tendencies, inclinations, and predilections would make it difficult for Section 103 to come into full effect; would make it difficult to achieve the two purposes of those who wrote it. They would, furthermore, make it seemingly impossible to bring about universal adoption of an approach that, when combined with the Section 103 obviousness test, minimizes the subjectivity encouraged by the A&P opinion.

That approach was one followed by Judge Learned Hand, who sat on the federal courts longer than anyone else (over fifty years, starting in 1909), and whose name is venerated throughout the legal profession.* Judge Hand suggested that it was necessary for a judge, in reaching a decision, to listen to and strongly consider *all evidence* regarding the entire milieu in which an invention was created. Although he participated in more than 200 patent cases, Judge Hand regarded himself as incompetent to decide them without strongly considering *all* the relevant evidence. His premise was that when there is other evidence (in addition to prior art) about patentability it should be used, but when "such evidence is not at hand, we [judges] are forced to fabricate a standard as best we can from our naive ignorance; but that is so unsatisfactory an expedient that resort to it should be as sparing as possible."

In the last of his patent cases, Hand said, "to judge on our own that this or that new assemblage of old factors was, or was not, 'obvious' is to substitute our ignorance for the acquaintance with the subject matter of those who are familiar with it." He then went on to discuss the other evidence, which he

*Only because Chief Justice Taft wrote that Hand was a "wild [Theodore] Roosevelt man and a progressive" did President Harding decline to put him on the Supreme Court. Later, in 1943, Franklin D. Roosevelt referred to Hand's "chronological age" as explanation for the failure to nominate him for the Court.

called "sign posts." These signposts were not originated by Hand; they came from earlier court decisions. But Hand followed them resolutely, and listed some of them:

- How long did the need for the invention exist?
- How many previous inventors had tried to find the solution?
- How long did the prior art contain the means to achieve the solution (that is, the elements necessary for the combination)?
- How soon was the invention recognized as the answer to the problem by those who used it?

The last point is similar to such significant questions as: How did people in the inventor's field react *at the time* the invention came along? Were they surprised? Did they first scoff at the invention and then change their minds and adopt it? (What "experts" say in court, after having been paid by lawyers to testify in favor of their clients, does not apply.) For example, the inventor of FM radio proposed to "find what the men skilled in the art, who had worked on the problem unsuccessfully, said or wrote at the time the invention was brought to their attention. That is the best judgment that can ever be obtained on the question of what was obvious when the invention was made."

Commercial success is another important signpost, but, of course, it would be naive to assume that anything should be patentable solely by virtue of its being commercially successful.

The obviousness test of Section 103 is the one that best promotes the use of signposts; it, in effect, tells a judge to weigh all the evidence, not merely to compare the invention with the prior art and come up with a subjective decision. Of course, even when signposts are used together with obviousness, there is still some subjectivity in the interpretation of the signposts and in the application of the test. To *ignore* the signposts, however, is to *deify* subjectivity.

The use of signposts whenever possible is the second of the two steps (the first being the application of Section 103 itself) that must be taken for there to be reasonable, minimally subjective separation of worthy patents from shadow ones. Nevertheless, most judges have been highly reluctant to accord signposts the weight of other evidence because the Supreme Court long ago indicated that signposts should be followed only in "close" (hard to decide) cases. This was referred to at the end of the Olga opinion, where the Court of Appeals said:

> While such secondary considerations as commercial success and long felt but unsolved needs in the field *constitute relevant evidence in a close case* on the issue of obviousness, once it is established by prior art references that the difference between the patents in suit and the prior art is not substantial enough . . . the patent cannot be sustained. (italics mine)

It seems curious that Olga's inventions were not even regarded as sufficiently good to make the question "close." The fact is that an appellate court, working in a desiccated atmosphere of transcripts, devoid of the human element present in a trial court, is deprived of a sense of the milieu that is so important in reaching a just decision. When such a court—or any court—downgrades the importance of signposts or refuses to consider them at all, it is as if, a leading patent attorney has said, "The court decided a negligence case solely on the basis of a road map without asking whether the light was red or green, whether the driver was drunk, etc."

The confusing nature and the lack of consistency of court opinions mentioning signposts was exemplified by the varying opinions of Justice Jackson. In overturning the lower court decisions sustaining the checkout counter patent, Jackson said:

> The Court of Appeals and the respondent both lean heavily on evidence that this device filled a long-felt want and has enjoyed commercial success. But commercial success without invention will not make patentability.

Less than two years earlier, however, in a case involving jewelry manufacturing, the same Justice had strongly dissented to the invalidation of a mechanical patent by declaring:

> Of course, commercial success will not fill any void in an invalid patent. But it may fill the void in our understanding of what the invention has meant to those whose livelihood, unlike our own, depends upon their knowledge of the art. . . . the only patent that is valid is one which this Court has not been able to get its hands on.

It is difficult to accept that one Justice voiced both these opinions within only two years. Chief Justice Hughes could scarcely have found a better example for his statement that court opinions are often rationalizations "for supporting our predilections."

Over the past century, there have been some cases in which the judges followed Judge Hand's signposts. Their line of thought has been something like this: This does not look to me like an invention worthy of a patent, but it did look like one to the people out there in the field, so I will hold the patent valid. A once-famous Supreme Court case of this type, which has since been largely ignored, related to the first battery enclosed (by its inventor, Herman Anthony) in a steel jacket in such a way as to prevent leakage and to lengthen shelf life.

The battery patent was owned by the then-small Ray-O-Vac Company of Madison, Wisconsin, and had allegedly been infringed by the Goodyear Tire and Rubber Company. In 1944 five Justices of the Supreme Court held that the patent was valid. They stated:

> *Viewed after the event,* the means Anthony adopted *seem simple* and such as *should have been obvious* to those who worked in the field, but this is not enough to negative invention. During a period of half a century, in which the use of flashlight batteries increased enormously, and the manufacturers of flashlight cells were conscious of the defects in them, no one devised a method of curing such defects. Once the method was discovered it commended itself to the public as evidenced by marked commercial success. These factors were entitled to weight in determining whether the improvement amounted to invention and should, *in a close case,* tip the scales in favor of patentability. (italics mine)

This was probably the most recent Supreme Court decision in which the sustaining of a patent resulted in a major change in market positions in a particular industry. Ray-O-Vac Company expanded rapidly, and today (as part of INCO ElectroEnergy Corporation) provides major competition for such giants as Union Carbide and its Eveready battery. This is also the only modern case in which the Supreme Court admitted, in effect, that while the invention before it did not, per se, seem deserving of a patent, such factors as long-felt need and commercial success could sway the balance.

It is such decisions that hold out hope for inventors in the future. By following Learned Hand's signposts, the Court proceeded toward an equitable judgment. It is also the way the Court proceeded in a 1966 case when it combined unobviousness with signposts and turned out an opinion so well conceived that it could have been written by Learned Hand himself. That case, however, was decided on the same day as two others that would—paradoxically—determine the ultimate fate of signposts and the obviousness section.

FOUR

One Up, Two Down in the Supreme Court

Who invented the steamboat? Was it Gerbert, the Marquis of Worchester, Newcomen, Savary, Papin, Fitch, Fulton? The fact is, that one new idea leads to another, that to a third, and so on through a course of time until some one, with whom no one of these ideas was original, combines all together, and produces what is justly called a new invention.

THOMAS JEFFERSON

The year 1952 was an encouraging one for inventors and patent lawyers. Enactment of the first entirely new patent law in 116 years raised hopes that the era of excessive strictness and confusion was over, that the oppressive A&P doctrines were dead, and that lawyers no longer had to worry about which test or tests would be applied to a client's invention. But as time passed, it became evident that many lawyers and courts were not paying nearly enough attention to the new law. The A&P case continued to be cited, and the lower courts often spoke in the pre-1952 language of "lack of invention" rather than "obviousness." Different courts ruled in different ways, and it was apparent that the Supreme Court would have to impose order. Patent lawyers waited for the Court to act. They and their clients waited a long fourteen years.

During those fourteen years, there was one bright spot. In 1960 Judge Learned Hand made a widely hailed statement, "There can be no doubt that the Act of 1952 meant to change the slow but steady drift of judicial decision that had been hostile to patents." This was exactly what patent owners wanted to hear, because a mere change in language obviously meant nothing if excessive severity continued to be the rule.

Then, in 1965, the news that the Supreme Court would hear not one but three patent-validity cases produced a surge of excitement. Meetings were held by various patent bar groups, and speculation was intense about what decisions and opinions the Court might reach.

The suspense ended on February 22, 1966, when newspapers revealed the score in rather extensive articles: alleged infringers 2, patent owners 1. In three decisions announced on the same day, the Court invalidated two patents for mechanical inventions while sustaining one for an electrochemical battery. The latter case will be described at some length because it was the first patent to be sustained by that Court since 1949, and it is a model, I believe, of how patents should be scrutinized by the courts. This case also turns out to be the very model of what most people think an inventor's story should be and an inspiration to anyone who has ever wondered what chance the little fellow can possibly have when he encounters what in this case *Time* magazine called "years of plain and fancy Government hornswoggling."

Bert Beats the Bureaucrats

Bert Nemo Adams was born in New Rochelle, New York, and was adopted by a man who later attempted to invent a new kind of soap. Whether the man was successful is not known, but his knack for working with his hands must have rubbed off on the boy because after graduating from high school in 1915 Adams took a job with General Electric as an apprentice mechanic. A year later, as war fever mounted, he obtained a berth in the Merchant Marine after trying unsuccessfully to enlist in the Navy.

Adams spent the war years ferrying troops to Italy, and he went back there in 1920 to study engineering at the Milan Didactic Institute. Four years later he returned to the United States to work for a Florida marine salvage firm until the stock market crash of 1929. During the Depression he "took any job which came along, automobile repair and radio and painting." By the Spring of 1939, he was living with his wife, Emma, not far from the World's Fair in Flushing Meadows, New York.

Despite his haphazard life, Bert Adams "wanted to set the world on fire," a lawyer recalls. An Italian company hired Adams to help put up an exhibit at the World's Fair, where he encountered a chemist working on a battery display. Adams plied him with questions about the chemical generation of electricity, an art that went back to 1800, when the famous Italian scientist Count Alessandro Volta made the first battery.

In one of his initial experiments, Volta had tried separating plates of copper and silver with cardboard and immersing them in water. After discovering that a continuous flow of electricity could be produced from such an assembly only after metallic salts had been added, he concluded that batteries required three good electrical conductors—the two electrodes and a liquid electrolyte. Plain water, being an insulator, was *not* a good electrolyte, Volta believed. For the next 140 years, Volta's discovery influenced the evolution of battery development. Adams' research, however, led him to think that it ought to be possible to make a water-activated battery. An important advantage would be that water, unlike the electrolytes used in Volta's "wet" cells, would not be corrosive.

Adams studied the formidable science of electrochemistry at the local public library and began looking for metals that would work better than carbon and zinc, which formed the electrodes in conventional batteries. He made shopping expeditions to chemical supply houses in lower Manhattan, and Emma helped carry his purchases home because he was in declining health. By experimenting in his small apartment, he eventually found that water would conduct a current when magnesium was used for one of the electrodes (the anode), and cuprous, silver, or gold chloride was used for the other (the cathode). He settled on cuprous chloride "because it is cheaper and has the same effect," but he was unable to obtain more than a feeble current.

Early in 1940, while puzzling over the problem of raising the current level high enough to make the battery practical, he had a heart attack and was hospitalized. Released in May, he resumed his experiments. With money running low, Emma took a job in a candy factory; but every evening she would, although bone tired after a day's work, assist him by carefully recording measurements made by a voltmeter. "He would work from morning to night, never giving up," she recalled. "He would even be lying in bed, and all of a sudden pop out of bed and go to work. I got used to that."

Adams melted the cuprous chloride in a crucible on the kitchen stove, shaped it in small, handmade molds, and assembled battery after battery in baby food jars. The molten black compound smelled so foul that Emma was often forced out of the room; eventually, the landlord asked them to move.

Bert Adams was a heavy smoker, the kind one invariably sees with a cigarette in his mouth, its ash growing longer and longer until it falls of its own weight. One night, while he was engrossed in his experiments, the ashes from his cigarette dropped into the melting cuprous chloride. Although he feared the mixture might be ruined, he had no place to dispose of it; so, hoping for the best, he continued cooking it and then fabricated his battery in the usual way.

This time the needle jumped when he connected the

meter—the current he had long been looking for was finally being generated. "I got it, I got it!" he yelled, startling Emma out of her sleep and causing her to think he might have injured himself, "because he would get burned quite a bit trying to lift the crucible; he was a very excitable person and worked very fast."

What Bert Adams had was a battery that would light a small bulb and produce a substantially constant voltage for the lifetime of the battery (roughly thirty hours at first). This was in striking contrast to conventional lead-acid batteries, in which voltage decreased as the battery operated. Adams' battery, furthermore, could sit on a shelf for a long time and then be activated merely by adding water. It was thus an excellent reserve battery capable of performing any number of emergency chores.

Exhilarated by his success, Adams set out to perfect his invention. Speculating that the carbon in the cigarette ash had served as a catalyst, he experimented with cathodes impregnated with charcoal, hard coal, powdered graphite, and even sugar. At night he placed the baby-jar batteries on the dresser so that he could watch them. Each had its own bulb, and Emma, a light sleeper, was periodically awakened both by the seven or eight small lights twinkling in the dark and by Bert jumping up to check them.

Eleven days after Pearl Harbor, Bert Adams applied for a patent on his battery, which was called the "Neutro Cell." It had taken two years to develop. He sold friends small shares in the invention (the patent for which was granted two years later) to help with the filing expenses. Then, having received a publication urging "the inventive minds of our country to transmit the results of their creative thought to that government agency charged with the duty of making them useful in this time of national emergency—The National Inventors Council," Adams wrote to the Council on January 7, 1942. The next day he addressed the Navy, and two days later the Army Signal Corps.

A four-day meeting in Washington between Adams and

technical experts at three different bureaus of the Navy resulted. Later, there were demonstrations before battery experts at the Army Signal Corps laboratories at Fort Monmouth, New Jersey. Adams promoted his battery as an auxiliary or emergency power source for submarines (it could be activated by sea water as well as fresh water), airplanes, bomb shelters, and portable radio transmitters. Because the battery produced heat (to near boiling point) while it was operating, it could power equipment in subzero cold, a feature that subsequently made it invaluable as a power source for the instruments carried by weather balloons into the stratosphere, where temperatures drop to 60 degrees below zero.

Adams made a number of trips to Fort Monmouth during the next four years, paying the expenses out of his own pocket and supplying batteries that the Army requested for tests. But despite repeated inquiries, Adams was unable to get a satisfactory report of the results, being told he would hear "in due time."

In January 1943, Adams was informed that some of his batteries had overheated and blown up. Although he believed the accidents had occurred because the batteries had been overloaded, the government later maintained that it was the "unpredictable exothermic reaction" of the Adams battery that was at fault, making it "impracticable for military purposes" and spurring the Army to undertake to improve it.

There had, in fact, been skepticism about Adams' battery from the start at both the Signal Corps and the National Bureau of Standards, where an expert felt that Adams was making "unusually large claims [for] high watt hour output per unit of weight." He found "far from convincing" Adams' graphs showing the battery's characteristics, and he concluded, "Until the inventor can present more convincing data about the performance of his cell, I see no reason to consider it further. If it is of any use, much development would be needed."

By November 1943, however, the Signal Corps seemed to

have changed its mind about the battery. A report that month by a Fort Monmouth investigator called the battery "feasible." The expert, noting that the battery "eliminates acid as the electrolyte and hence the evolution of dangerous fumes is avoided," pointed out, "it can be easily constructed and maintained provided the internal resistance of the water electrolyte is partly overcome." The government subsequently contracted with various companies for procurement of a water-activated, magnesium–cuprous chloride battery. At least 1 million were eventually produced for use with weather balloons and survival gear for military fliers.

Adams, persisting with his efforts to perfect the battery, returned to Fort Monmouth in 1948 to deliver yet another of his cells. The Signal Corps, however, as the Supreme Court later noted with surprise, "did not notify Adams of its changed views nor of the use to which it was putting his device, despite his repeated requests." In 1953 the U.S. Government obtained a patent for a magnesium–cuprous chloride battery embodying the modifications of two of the Fort Monmouth scientists whom Adams had met in the course of demonstrating his invention.

It was not until 1955, when Adams first saw a water-activated battery made for the government by a major manufacturer, that he discovered what had become of his invention. Furious, he sought a lawyer and had the good fortune to end up in the office of John A. Reilly, today a partner in the New York City law firm of Curtis, Morris and Safford. Reilly, a genial, expansive man, is regarded with awe by students of the case.

One of Reilly's first moves was to solicit the aid of Lewis Schott, an investor who Reilly felt might be sufficiently aroused by Adams' plight to advance the funds necessary to pursue the case. Schott was outraged. "I thought they had *shmeissed* this fellow," he recalls. "All I wanted was for Adams to get the recognition that he deserved. And that's all he ever wanted." Before the case drew to its close, Schott had provided nearly $200,000, a sum far greater than he had contemplated.

In 1957 Reilly petitioned the Army to compensate Adams for use of the invention. At that time a settlement could have been negotiated for as little as $25,000, Schott believes. At one meeting with Army officials, however, Schott recounts, "Not only did they deny that they had stolen the patent, but, more important, a general said that even if everything we said was true, it wasn't worth $25,000. Then I knew the battle was on."

The Army's rejection of Reilly's petition came on the heels of a scientific paper presented at a research and development conference the previous year. The author had hailed water-activated battery systems as "among the relatively new power sources which have become extremely important to the Military during the past decade. Their inherent characteristics are such that their use will expand as the Armed Forces continue to utilize more and more equipment to increase the efficiency of the soldier. . . . [They have] brought about developments which would otherwise have been technically or economically impractical."

On May 2, 1960, only seven weeks before the Adams patent was due to expire, a lawsuit was filed by Reilly in the United States Court of Claims, which was the only court where the U.S. Government could—at that time—be sued for patent infringement. (The cases were tried by individual trial commissioners whose decisions and opinions were reviewed—and usually followed—by the Court itself.) The plaintiffs included Bert and Emma Adams, friends who had advanced money in the invention's early days, and Schott. The case was referred to one of the trial commissioners, Donald E. Lane, a highly experienced former patent lawyer who later became a member of the Court of Customs and Patent Appeals.

Much of the argument before Commissioner Lane revolved around the meaning of the word "fused" as it pertained to the cuprous chloride electrode in most of Adams' eleven patent claims. The government maintained that "fused" implied that temperatures of 400–500°C were required to form the electrode, as the Adams patent specification described. Because none of the government batteries were made in this fashion, defense counsel argued there had been no infringement.

Reilly, on the other hand, contended that "fused" merely meant "unified," and it was his view that ultimately prevailed. Commissioner Lane ruled that the patent was valid and that "the invention covered by said patent was used by defendant without authorization of the patent owners, and that plaintiffs are entitled to recover reasonable and entire compensation for such unauthorized use, and judgment will be entered to that effect."

The government requested the Supreme Court to rule on the patent's validity but left unchallenged the lower court's finding that there had been infringement. After the usual briefs, answering briefs, and delays, Reilly arrived at the Supreme Court on a Monday afternoon in October 1965. He listened to oral arguments of other cases for the next three and a half days in order to get a feeling for the different Justices and their approaches. On the day the three patent cases were to be argued, he was scheduled to be last. He recalls that as he sat in the venerable courtroom packed with patent lawyers, the nature of the other two cases—which dealt with improvements in existing mechanical products—gave him the increasingly queasy feeling that this would not be a good day for inventors.

Reilly, however, stood up to present an argument that was called "a classic of effectiveness" by one of Justice Tom C. Clark's law clerks, who worked on the first draft of the Court's opinion.* Reilly pointed out, "When Adams took his battery a

*So-called law clerks are, in fact, aides to the Justices. They are culled from the top law school graduates in the country the previous year and generally serve only one term.

month after Pearl Harbor and went to Fort Monmouth, as he was directed to do, and displayed it to the assembled battery experts . . . they looked at the battery working in front of them with the steam coming out of it and told Adams, 'It can't work.' That's the battery the Government says is obvious." Bert Adams, said Reilly, was no scientist. "He didn't know that what he was doing was impossible. He didn't know that the combination he put together was ridiculous."

Reilly demolished a British patent that the government presented as an exhibit of the prior art. Issued in 1880 to a Russian named Skrivanoff, who lived in Great Britain, it was a combination of magnesium and cuprous chloride but with numerous additional substances, including sulfuric acid and phosphorus powder. Reilly told how he had hired an expert to construct a battery according to the Skrivanoff patent. But when the man tried to build it, the result was an explosion that blew the vent out of his laboratory. The story of the explosion created considerable amusement among the Justices.

Of great importance was the fact that the invention could be dramatically demonstrated before the Justices' eyes. Justice Clark's law clerk recalls:

> John brought a working model of the battery . . . to the argument. The model was an unpretentious glass of water, electrolyte, chemicals and a Rube Goldberg combination of flimsy wire electrodes and lightbulb. . . . As he argued, he fidgeted with his model. I recall Mr. Justice Black cracking a joke (perhaps not intentionally) by asking whether the contraption was going to explode. John assured him it was safe. I didn't know it at the time, but John had experimented with different temperatures of electrolyte to control the rate at which current was to be generated. On that day, he used a colder solution. The result was that when John's oral argument had ended, his battery had not. His opponent had to argue as all the

while John's (or Mr. Adams's) battery powered that tiny, but attention-gathering, lightbulb. Indeed, the demonstration sputtered on into the next case, quietly yet eloquently mocking the Army's lack of confidence.

On February 21, 1966, the Supreme Court pronounced Bert Adams' victory in his twenty-five-year struggle with two words: "We affirm." The opinion was written by Justice Clark, who called the Adams battery novel, unobvious, and therefore valid. In considering the question of obviousness, Clark pointed out how "noted experts expressed disbelief" and how "several of the same experts subsequently recognized the significance of the Adams invention, some even patenting improvements on the same system."

Reilly's demonstration and the tremendously human story remained so vivid in Justice Clark's mind that, on the same day the Court's opinion was issued, he related the tale of Adams' invention to a group of visiting school children.

The decision made Bert Adams something of a hero, and a number of newspapers and magazines paid tribute to the inventor, who was now sixty-six years old, ailing, and living with his wife in Yuma, Arizona. A photograph in *Time* magazine's "The Law" section showed a stocky man with a sport jacket, western string tie, and horn-rimmed glasses smiling broadly under the headline, "How Bert Beat the Bureaucrats." The *Christian Science Monitor* reported that he said: "I knew that justice always wins out. I had faith that it would, and it did."

But it would take another two years before the actual cash settlement was negotiated, and then only after Michigan's Democratic Senator Philip Hart introduced a bill to extend the Adams patent ten years. Said Senator Hart: "It seems very clear to me as it did to the Court that someone in the Government has been deliberately unmindful of the rights of this man. Simple justice needs to be done. . . ." Although the bill was not passed, the government settled with the plaintiffs shortly afterward for $2.5 million, of which $500,000 went to

Bert Adams.* By then Adams had less than a year to live. But at least he had the satisfaction of "beating the bureaucrats," and if he were alive today he would undoubtedly draw even greater satisfaction from knowing that every day a few hundred updated versions of the battery he cooked up on his stove that long ago August night still soar into the stratosphere on weather balloons. As Reilly impressed on the Supreme Court: "If you have flown . . . in the last ten years, your course, altitude and safety were determined by information radioed down to earth by the power of Adams batteries."

Why Bert Adams Won

The Supreme Court Justices decide their cases at periodic conferences where only they are present, and secrecy is so strict that the junior Justice is used to carry messages to the outside world. The conference-room door has a core of sand to soundproof the room. Votes are recorded in notebooks kept under lock. Consequently, we cannot know with certainty just how much difficulty the Justices had in deciding the fate of Bert Adams' patent. (One Justice, Byron R. White, dissented, while Justice Abe Fortas abstained.) However, it should not have been hard for the Court to conclude that Adams' invention was unobvious. First, the subject matter was chemical, specifically electrochemical, and thus not subject to the discrimination against mechanical devices that was spawned by the A&P decision. Second, the invention was such that John Reilly was able to make his dramatic technical demonstration in the courtroom. And, he impressed the Court with statements such as "The Adams battery characteristics are opposite, in most cases, to every prior art battery—140 years of battery research—this is the first different battery since Volta invented the battery in 1800." Furthermore, Reilly could tell the Court that to this day "no one really understands" why the Adams battery works as it does. Then there was the matter of

*The great majority of the remainder of the money went to eight persons who were co-plaintiffs with Adams.

the luck Reilly had in being able to tell the Court how Skrivanoff's battery had blown up when Reilly's expert tried to build it. Last, but probably most important, there was the solid "signpost" evidence—the skepticism among experts, the demonstrable need for the battery, and its widespread use.

The Court's (Justice Clark's) opinion in the Adams case was a classic of lucidity. Probing the complex subject in its entirety, the opinion analyzed such signposts as the skepticism of the experts at the time. The opinion did *not* make the typical mistake of saying that the oldness of the individual components indicated obviousness. Said the Court: "It begs the question . . . to state merely that magnesium and cuprous chloride were individually known battery components. If such a combination is novel, *the issue is whether bringing them together as taught by Adams was obvious in the light of the prior art.*" (italics mine) In making this statement the Court was acknowledging one of the keys to successful inventing: the ability of the inventor to select from a universe of prior art precisely those components or concepts that are necessary to achieve something new and unique. After all, any reader of this book can make an Adams battery once someone shows him how.

One knowledgeable commentator noted bitingly that this was the first patent to be sustained by the Supreme Court in an extraordinarily long time:

> [Reilly] fought the strong trend of 25 years and persuaded the Supreme Court to say that there is such a thing as patentable invention after all. But, unfortunately, the test the Supreme Court handed down for the circumstances of patentable invention requires a client who is an individual inventor, who comes to the United States Government with his invention, who is lied to by the Government officials, and wherein Government agencies use the invention without his knowing and make it impossible for him to find out what they are doing. Then he [Reilly]

may provoke the Supreme Court to concede, with the aid of the wizardry of a technical demonstration in front of the Court, that perhaps there might be some room still for a patent to emerge.

Bad News for Inventors

Unfortunately for inventors, the opinion in the Adams case has *not* had a great effect on the lower courts. Rather, it has turned out to be, as Reilly lamented years later, "a much-neglected precedent." Reilly discovered that between 1966 and 1976 Adams was cited in roughly 20 percent of the 581 appeals court validity cases and in less than 10 percent of 979 district court cases. (These figures relate only to published decisions; many are not published.) On the other hand, the Supreme Court decisions invalidating the two mechanical patents, on the same day as the Adams case, were cited in a vast majority of patent cases during the next decade.

One of the two mechanical patents related to a plow invented by William T. Graham and allegedly infringed by the giant John Deere Company of Kansas City. The other involved an insecticide bottle spray-cap (sprayer) developed by Baxter I. Scoggin Jr. The spray-cap patent was assigned to the relatively small Cook Chemical Company of Kansas City and was allegedly infringed by the larger Calmar Company and the renowned Colgate-Palmolive Company. Both inventions were "simple" and not "technological." As with the checkout counter of A&P, anybody looking at them would tend to say, "big deal!" (It was a stale aphorism among lawyers for patent owners that if the judge understands it, you are dead.)

The Supreme Court adopted the highly unusual procedure of discussing the two mechanical cases in a single opinion, with a common preface for both. The preface opened with a review of patent history and the background of the 1952 Patent Act, in the course of which the Court acknowledged that the word "invention" was essentially undefinable; consequently, "Its use as a label brought about a large variety of opinions as to its meaning. . . ." The Court then addressed the principal reason for agreeing to hear the three patent cases: to determine precisely what effect Congress had intended the 1952 Act, particularly the obviousness section, to have on the issue of patentability. The Court's answers, inventors and their lawyers thought, made sad reading.

First, the Court depreciated the value of signposts by calling such considerations "secondary":

> Such *secondary* considerations as commercial success, long felt but unsolved needs, failure of others, etc., *might* be utilized to give light to the circumstances surrounding the origin of the subject matter sought to be patented. As indicia of obviousness or nonobviousness, these inquiries *may* have relevancy. (italics mine)

Second, the Court denied that Congress had intended—as many persons (including Judge Learned Hand) believed it had—to make it less difficult to get a valid patent. The Court said that the obviousness section "was not intended by Congress to change the general level of patentable invention."

Third, the Court emphasized that the obviousness section was to be applied with "strictness." Said the Court: ". . . it bears repeating that we find no change in the general strictness with which the overall test is to be applied."

Finally, the Court's opinion was insufficiently definite and emphatic about whether the statutory obviousness test was to be the only one, was to replace the entire maze of judge-made patentability tests. There were statements in the opinion which *should* have been interpreted as meaning that obviousness is the sole test. However, because the judge-made tests, especially the strange tests generated by the A&P case, were cast in concrete in the minds of many lawyers and judges, it would have taken a verbal "hammer blow" to knock them completely out. That hammer blow was not delivered.

It must be emphasized that the Court had grounds for not fully following the intentions of the writers of the obviousness section (that is, to moderate the extreme degree of strictness and to get rid of all judge-made tests). Unfortunately, those who worked on the section did not make their purposes sufficiently clear at the time the law was passed, either in the section itself or in writings (called Reviser's Notes) relating to the legislation. (In writings and speeches made subsequent to passage of the law, their intentions were made abundantly clear. But when a court interprets a law—gets at what is called the "Congressional intent"—it normally disregards as irrelevant things said or written after the law was passed.)

After laying down its unsettling and inconclusive guidelines, the Court used them to invalidate the patents for Graham's plow and Scoggin's spray-cap. Graham, the owner of a small Amarillo, Texas, farm equipment company, had patented an improvement to a type of plow having a chisel attached to the rear end of the shank. As originally constructed, chisel plows were unsatisfactory in rocky soil because tremendous shocks could be transmitted when the chisel struck a buried rock, causing the shank to break. Graham added a spring clamp that allowed the shank to be pushed upward when the chisel hit a rock, thus absorbing the shock. In 1953, he patented a modification of the clamp, and it was this modification that became the subject of litigation. The Supreme Court was not impressed by Graham's modification, and decided that it would have been obvious.

The Scoggin patent involved a leakproof sprayer (spray-cap) integrated with the bottle cap and bottle of insecticide. This may not sound like much of an achievement, but in the days before aerosols, insecticides were sold in bottles capped with finger-operated sprayers. Because of their low viscosity, the insecticides tended to leak when the plungers were held down during shipping and storage. As a result, the spray pumps were not shipped in the bottles at all but were hung on the side by paper attachments. There were costly problems with this method, so Scoggin (who was Cook's vice president in charge of production) "married" the sprayer with the bottle in a way that was suitable for mass production. Cook then produced 15 million of the units.

Although an alleged infringer, Calmar, had been virtually forced by its customers to use a similar cap after its own research on other types of solutions had not resulted in customer acceptance, Calmar attacked Scoggin's spray-cap invention as being a "tinker's rearrangement" and a "transitory, petty refinement of a gadget."

The obviousness section of the 1952 Act says that a patent may not be obtained "if the *differences* between the subject matter sought to be patented and the prior art are such that the subject matter *as a whole* would have been obvious. . . ." (italics mine). The lower courts had made this comparison and held the sprayer patent valid and infringed. The Court of Appeals for the Eighth Circuit declared that the sprayer "meets the exacting standard required for a combination of old elements to rise to the level of patentable invention by fulfilling the long-felt need with an economical, efficient, utilitarian apparatus which achieved novel results and immediate commercial success." In essence, the Court of Appeals applied Hand's signposts instead of its own reactions.

The Supreme Court, however, looked primarily at the size and type of *differences* between the subject matter and the prior art, without delving adequately into the question of whether the subject matter *as a whole* would have been obvious. Essentially, it dissected the invention into its component parts and decided that it "rests upon exceedingly small and quite nontechnical mechanical differences in a device which was old in the art." With regard to Judge Hand's signposts, the Court concluded that commercial success and long-felt need "do not, in the circumstances of this case, tip the scales of patentability."

Patent lawyers, nevertheless, did not greet the rulings on the two mechanical cases (which are usually—because of their single preface—collectively called "Graham" after the plow inventor) with anything like the scorn they reserve for A&P. Relieved at having one patent sustained by the Supreme Court after a seventeen-year drought, they could not know that Graham would overshadow Adams in the minds of lower court judges. The Graham opinion, furthermore, had the saving graces of omitting such catchwords as whole-must-exceed-sum-of-parts and of employing obviousness as the test of patentability. Considering that there had been 116 years of assorted "lack of invention" verbiage, this was a major advance. Lower courts were directed to resolve the obviousness question by what has since come to be called the "Graham approach":

> Under [Section] 103, the scope and content of the prior art are to be determined; differences between the prior art and the claims at issue are to be ascertained; and the level of ordinary skill in the pertinent art resolved. Against this background, the obviousness or non-obviousness of the subject matter is determined.*

It is perplexing to find the Graham opinion in the same package as Adams. How does one explain the highly unfortunate and (as events would later prove) damaging language that effectively undermined the very signposts that the Court followed so carefully in Adams? It can only be speculated that the opinion was originally written somewhat differently but that as it circulated among the Justices it was modified to meet the objections of those who were unimpressed with seemingly simple mechanical inventions. In the tug-of-war that could

---

*Despite the Graham opinion, there are still some who fail to realize that—since passage of the Patent Act of 1952—there are *two* kinds of "inventions": those that are patentable and those that are not. Before 1952, the word "invention" was used to mean "patentable invention." Thus, before 1952, if something was not considered to be an "invention" under the tangle of tests, then it was not patentable, and vice versa.

conceivably have taken place over Graham, it is possible that words such as "secondary" were added without full appreciation of how inconsistent they were with the Adams opinion.

After these three cases, the legal weather grew heavier for inventors. Many lower courts tended to ignore the Adams signposts and to pounce on the words "secondary" and "may" in Graham. By not expressly and absolutely slamming the door on A&P, and by emphasizing strictness, Graham spelled trouble for inventors all over the United States.

FIVE

Damage in the Lower Courts

My guess is that well over 80 percent of the human race goes through life without having a single original thought.

H. L. MENCKEN

Because few patent cases ever arrive in the Supreme Court, the number of inventors *directly* affected by particular Supreme Court decisions has been miniscule. It is in the lower courts, where the precedents established by the Supreme Court are applied, that the real action in patent litigation takes place.

That Graham strongly influenced the thinking of lower-court judges is evident from the statements a few of them have made. For example, shortly after Graham, Circuit Judge Harry A. Blackmun (now a U.S. Supreme Court Justice) declared: "as in other post-Graham decisions . . . we conclude that Graham *necessitates a holding of invalidity* of the patent here, and that the baffling problems which confronted the inventor in Adams prevent that case from having any factual correspondence with or control over the present controversy." (italics mine) Judge Blackmun was then on the Eighth Circuit Court of Appeals, which had long been tough on inventors but became even stricter after Graham. In fact, in the nineteen test-of-validity cases decided by that Court of Appeals during the six years after the Graham opinion, only one patent was sustained. The Eighth Circuit became by far the most tough-on-patents circuit in the United States.

Almost every patent lawsuit involves two main questions: (1) Is the patent valid? (2) Is the patent infringed by the particular construction that the alleged infringer is making, using, or selling? To win in court, patent-owning plaintiffs must win on both of these issues. Usually, such plaintiffs won't sue an alleged infringer unless their case relative to the infringement issue is reasonably good. The reason? Predicting a court's decision on infringement is much less difficult than guessing how it will handle validity, and going to court when there is a strong chance that there will be a finding of noninfringement is too risky and expensive.

So, when inventors go to court today, they commonly find themselves in the same predicament as Olga Erteszek; instead of concentrating on whether or not their patent was infringed, they are battling primarily over whether the patent was valid to begin with. And here the crucial question is whether the invention has passed the tests of patentability (such as unobviousness). Graham helped defeat Olga in the Second Circuit Court of Appeals, which elected to use only obviousness as its test and used Graham's "secondary" language regarding such considerations as commercial success. However, Graham has played a more obvious role in invalidating a great many of the patents litigated in the lower courts after 1966. Here is one such case.

Murray Crosby Creates Stereo FM Radio

Nineteen hundred sixty-one was one of those legendary boom years on Wall Street. Hot new issues paced a frenzied over-the-counter market to new highs, and all Stock Exchange issues posted a gain of $70 billion for the year. But on April 20, the Federal Communications Commission issued a ruling from which a little electronics company in Syosset, New York, never recovered.

That company was Crosby-Teletronics. Its chairman, fifty-eight-year-old Murray Grimshaw Crosby, was a highly regarded and productive inventor who bore a striking resemblance to that quintessential 1950s figure, Dwight D. Eisenhower. Crosby grew up sixty miles northwest of Madison, Wisconsin, in the small railroad town of Elroy, where his father worked in the railroad company office and his mother taught piano. Constantly experimenting with electricity, he rigged a telegraph to his best friend's house, and operated ham radio station 90AX from his attic workshop. But he also found time to assemble (and wreck) his own motorcycle, to play saxophone in the Elroy High School orchestra, and to serve as humor editor of the "Turn Table." Ironically, in light of his future travail as an entrepreneur, his high school motto was "All I need is plenty of time."

After graduating from the University of Wisconsin in 1927 with a bachelor's degree in electrical engineering, he went to work at RCA's transoceanic communications laboratories far out on New York's Long Island. Over the next seventeen years, Crosby acquired over 100 patents for RCA, mostly in the emerging field of FM (frequency modulation) communications. So important did Crosby become to the electronics giant that, after he left in 1944 to join a small consulting firm, RCA paid him a handsome retainer to appear in court as an expert witness whenever required. Earnings from this arrangement helped him start Crosby Laboratories in 1948. There he embarked on the work that led to the commercial FM stereo radio millions of people take so much for granted today. Before going into that story, however, we will look briefly at the invention of FM itself by a man who was both Crosby's friend and one-time courtroom adversary, Major Edwin Howard Armstrong.

"The Major," as Armstrong was called, was one of the

world's truly great inventors, and FM radio is probably the best remembered of his inventions. Until Armstrong came on the scene, radio and static were practically synonymous, the result of the standard practice of putting sound on a radio wave by causing the amplitude (strength) of the wave to vary in direct proportion to the intensity of the sound (amplitude modulation, or AM). After spending years pondering the problem, Armstrong realized that AM made static inevitable because it has the same characteristics as the radio signals themselves—the same strength, frequency, and pitch. Thus, noise from the transmitter's electronic tubes, from nearby auto ignitions, from lightning, or from countless other sources was as likely as Beethoven or Brahms to issue forth from those old wooden radio cabinets.

Armstrong saw that it would be necessary to devise an entirely different means of imposing the sound on the wave. After years of experimenting, he applied for a patent (granted in 1933) on a method of varying the frequency of the radio wave rather than its amplitude. With frequency modulation, Armstrong revolutionized radio. Not only did FM overcome static, but it vastly increased fidelity because it operated at higher frequencies than AM and had less noise in its circuitry, thus making it possible to transmit the full range of the sound.

An awed *Fortune* writer captured the revolutionary impact of FM with these excited words:

> When you move to the porch the illusion of a piano playing in the next room is so strong that it requires a mental effort to remember that you are listening to a radio. A steel guitar solo comes over without any "mushing" of its peculiar, twangy vibrato. An announcer whispers into the microphone, and you start as if someone had whispered at your elbow. He strikes a match to light a cigarette. The match flame crackles lightly and a soft breath blows it out. He pours himself a drink, and the liquid realism of it makes your eyes turn involuntarily to

your host's whiskey decanter. The announcer taps two Indian elephant bells with rapid dissonants in the upper register. The shimmering afterglow hangs in the room with uncanny, lambent clarity. In the pauses between numbers the station is so silent that it is impossible to tell whether it is on or off the air. Sound and music glow against a static-free, dead-silent background.

But one thing was missing: there was no way to capture the full realism of the concert hall, with its violins on the left, cellos on the right, and percussion in the rear—what today we call a stereophonic effect. Leopold Stokowski had provided a remote audience with a vivid demonstration of this effect the same year Armstrong's patent was issued. The famed conductor placed three microphones before the orchestra in a Philadelphia studio and had the music transmitted over wires to three loudspeakers in Washington's Constitution Hall, where several hundred people listened raptly.

About that same time, in the futuristic kaleidoscope of the Chicago World's Fair, Murray Crosby marveled at a Bell Telephone Laboratories demonstration of "binaural" sound. Ushered into a second floor auditorium, in the sickle-shaped group of white, red, yellow, gold, blue, and silver buildings devoted to progress in the electrical arts, he and other tourists took their seats around the outside of a small glass-enclosed room. Inside, a tailor's dummy named Oscar stood with a microphone in each ear. The audience was instructed to pick up a pair of telephone receivers and lean forward on the rail in front of them, holding the receivers firmly to their ears.

Thirty-three years later, in court, Murray Crosby described the "realistic and startling" event that followed. A door opened in the back of the little room and the master of ceremonies stepped inside. He walked over to Oscar and whispered in his ear that he should please move over. At that moment "the whole audience lifted up," Crosby recalled, "and moved over to the next seat, because it was such a natural reproduction and it came in the right ear, just as though the person next to you had spoken into that ear and asked you to move over." In May 1933 a Bell technical journal predicted: "The system will furnish to those who use it dramatic proof that there is no longer any limitation, except expense, to the acoustic fidelity which electrical transmission systems can achieve."

Oscar continued to fascinate audiences throughout the 1930s and even put in an appearance at the 1939 World's Fair in New York City. During the same period, from his basement laboratory in Columbia University's Philosophy Hall, Major Armstrong began his long, lonely battle to bring FM broadcasting to commercial reality, a struggle that led him to spend a good many of his last days in a federal courtroom trying to enforce his patents against infringers.

Soon after Armstrong's first FM patents were issued, he began experimenting with FM mutiplexing—the "piggy-backing," as it came to be called, of a subcarrier channel on top of the main carrier. Multiplexing enabled background music, stock reports, and other specialized information to be transmitted along with the main channel in a way that was audible only to those listeners equipped with special receivers. In later years the technique became widely used to pipe background music into stores and restaurants.

FM multiplexing laid the foundation for stereo broadcasting, although stereo was not something that interested Armstrong until many years later. Some people accused him of having a tin ear, but Armstrong's long-time co-worker, John Bose, claimed in a 1966 radio panel discussion that "ideally, what he [Armstrong] wanted to do was play good music on the main channel all day long without any commercial interruptions whatsoever. However, to support this, he would need some economic way to pay for it. His solution was to use the subchannels for perhaps broadcasting of news, storecasting—what have you—to pay for the main channel program. And for this purpose our technical goals were very severe."

And so, from 1934 until 1954, when he took his own life, Edwin Armstrong devoted his prodigious talents and energies to developing the expensive and sophisticated equipment necessary to perfect FM and its transmission of subchannels. The first regularly scheduled FM music program was heard in 1939 by an estimated twenty-five possessors of FM sets in the New York area when a specially installed telephone line carried Haydn's Symphony No. 100 and Tchaikovsky's "Francesca da Rimini" from the studio of WQXR, the radio station of *The New York Times*, to Armstrong's experimental FM transmitter overlooking the Hudson River at Alpine, New Jersey.

By the early 1950s, an emerging audience of "hi fi" buffs was beginning to sense the potential of stereo broadcasting. Different schemes were tried, but perhaps the best remembered was one by WQXR, which the station's co-founder, Elliott Sanger, recalls as being more or less a stunt. A microphone at one side of the studio orchestra fed sound to an AM transmitter, and a microphone on the other side fed an FM transmitter. Listeners at home were instructed (by means of diagrams in *The Times*) to sit midway between their FM set and their AM radio. If both receivers were properly matched, and if the reception from both transmitters was equally good, they would hear a stereo concert. These conditions were practically impossible to meet, and even though a number of stations copied the scheme, there seemed little future for such an arrangement. Because the AM channel was inferior to the FM channel to begin with, that side of the orchestra would always sound worse than the FM side. Moreover, if you had only one receiver at home, you would hear only half the orchestra. In effect, it would be as if you were permanently condemned to a seat along the far wall of the concert hall.

What was needed before FM stereo radio could become a product for more than a core of die-hard hi-fi buffs was compatibility; that is, listeners with one ordinary *monophonic* FM receiver should still be able to hear *all* of an FM broadcast.

They should, in effect, be given a seat in the center of the concert hall. This compatibility was one of the important things that Murray Crosby was able to achieve.

During 1951 and 1952, Crosby was working on an FM multiplexing system for the Navy. Stereo records were just coming on the market, and Crosby, who loved popular music, thought that by integrating an FM multiplex system with stereophonic inputs he might be able to put a monaural program on the main channel and stereo on the subcarrier.

One morning Crosby suggested to Barton Coffman, his chief engineer, that they try to electronically combine signals from the left and right channels in such a way as to produce two new signals, one the algebraic sum of the two signals, the other the difference. The sum signal would be broadcast on the main carrier, the difference signal on the subcarrier. Consequently, a listener with an ordinary monophonic receiver would receive only the sum signal and thus would hear a balanced mixture of the two channels, just as if the program were being broadcast in the conventional manner. In a stereo receiver, however, the two signals would be directed through electronic circuits that first added them together and then subtracted them. The left and right channel signals would be reconstructed separately and fed to spaced-apart speakers, creating the stereophonic effect.

Coffman remembers that they put some circuits together, played some music, and found that the system worked. But there was still a long way to go. During the next few years, they strove to perfect the system, experimenting with different circuits in order to obtain the maximum possible stereo separation and fidelity. With Crosby's invention they also eliminated one of the main problems of FM stereo broadcasting: that one channel would invariably be noisier than the other. Crosby's system balanced the noise so that it was, in effect, cancelled out for the listener.

Crosby filed his patent application on April 21, 1953. In June he presented a paper on the invention before a session on binaural broadcasting at a meeting of the American Institute of Electrical Engineers in Atlantic City. Other speakers included none other than Major Armstrong, who reported that his Alpine, New Jersey, transmitter was operating with recorded music on one channel and a Washington, D.C. musical program on the other, in order to determine whether or not there was any interference. The station was also broadcasting "short binaural programs."

Crosby's paper was widely publicized, and his laboratory became a mecca for the top technical people in this field. During the next few years he gave at least fifty demonstrations to visitors, including representatives from such leading corporations as General Electric, Sylvania, NBC, and RCA. By 1961, Admiral and some thirty smaller companies had taken licenses to operate under Crosby's patent, which had been issued in 1958.

Meanwhile, as the stereo record business grew, it became possible to discern the outlines of a significant market for FM stereo receivers, and a number of companies—some giants in the industry—jumped into the race to develop a commercial system. During the summer of 1960, six systems were entered in field trials that an electronics industry committee conducted for the Federal Communications Commission. By authorizing the conditions under which commercial FM stereo broadcasting could occur, the FCC had the power to make or break any company relying as heavily on winning that competition as Crosby (now Crosby-Teletronics) was; for royalties were already starting to come in from licenses, and Crosby executives were beginning to envision Crosby-Teletronics as a household name under which they could conceivably manufacture and sell millions of their own receivers.

The crucial field trials were conducted for several days, from midnight until 6 A.M., over Pittsburgh's famous KDKA, the first commercial broadcasting station in the United States. (It had carried the election night news of Harding's victory over Cox in 1920.) Crosby felt that his system had come off well in the trials, and he was right. After receiving the committee's report, the FCC narrowed its final selection down to Crosby's and just one other—a composite of two almost identical systems from Zenith and General Electric, referred to as System 4-4A. The Crosby-Teletronics team was so certain it would win the competition that it is easy to imagine the team's shock when the FCC's decision was handed down on April 19, 1961, stating in the last column, "We find, therefore, that the public interest would best be served by the adoption of System 4-4A."

The "therefore" came about because the FCC found that 4-4A had "the clearly decisive advantage of being able to provide stereophonic broadcast service with negligible effect on the monophonic listener and that the correlative disadvantage of System 1 [Crosby's] is its detrimental effect on the monophonic listener." This statement referred to the fact that Crosby's system, where both the carrier and subcarrier were frequency modulated, had a greater loss in signal-to-noise ratio for monophonic reception and lesser loss for stereo, while just the opposite was true for the GE-Zenith system, in which the subcarrier was amplitude modulated. The FCC was saying, in other words, that it didn't want to penalize the monophonic listener. But in a footnote, the FCC also noted that it was "impressed" by 4-4A's apparent lower cost, comparative freedom from distortion, and "the fact that its use does not ipso facto displace SCA operation."

To this day Crosby supporters believe that the Subsidiary Communication Authorization (SCA) of 1955 was the real fly in the ointment. Under SCA, FM stations were allowed to furnish private subscribers with background music and other services by means of multiplex transmission on an additional subcarrier. Given the fragile state of FM broadcasting in those days, this service may well have been all that stood in the way of bankruptcy for many of the 200 stations that were estimated to be providing this service for restaurants and factories. The difficulty with Crosby's system was that because it was totally wideband FM it left no room in its bandwidth for the SCA service. Consequently, the fate of the stations dependent upon SCA for revenues was one of the factors the FCC had to consider in reaching its decision.

The decision broke Crosby's heart, says a long-time friend and associate. It also broke the pocketbooks of numerous investors who watched the stock plummet on the over-the-counter market that day, recovering only enough to end the week at 6¼ bid as against 9⅛ bid the week before. But far worse was the effect on the little company itself. A GE executive hailed FM stereo as "the most stimulating development in the radio industry in the past fifteen years," and both GE and Zenith announced they would begin selling receivers as quickly as possible. Crosby-Teletronics' royalty income dried up, and in November it filed for bankruptcy. A trustee was appointed in July 1962, and soon afterward the company grasped at its last chance for survival: it sued GE for allegedly infringing the Crosby patent.

Crosby Takes GE to Court

General Electric did not contest infringement when the trial opened in the U.S. District Court for the Eastern District of New York on a Monday morning in September 1966. (This does not mean that infringement was admitted; it means that General Electric chose not to fight that issue.) Rather, it introduced eleven pieces of prior art to demonstrate that the patent was invalid. One was a U.S. patent issued on December 28, 1954, to W. W. Boelens and J. J. Zaalberg van Zelst, based on a 1951 application originally filed in the Netherlands, where the inventors lived. The Boelens patent made no reference to FM broadcasting. It disclosed a system, using a sum-and-difference approach, for transmission and reception of stereophonic signals for AM broadcasting. The sum signal was sent out intact, but the difference signal was greatly modified so that considerable imbalance, accompanied by a lack of full fidelity, resulted.

Crosby had seen the Boelens patent described in the U.S. Patent Office *Official Gazette* shortly after it was issued in

1954. He had obtained a copy, given it to his patent attorney, and suggested that it be sent to the Patent Office with "our argument on it." This was done, and the Crosby claims were amended to emphasize the elements of Crosby's invention that Boelens did not include.

There had, therefore, been a "shift" in Crosby's position after the Boelens patent came to light. Now in many fields of endeavor there is no stigma attached to people who change their position when they learn of a new condition. But in patent law, particularly in recent decades, such a shift can be (and probably was here) a major factor in bringing about the downfall of an entire patent. (The situation is somewhat analogous to one of the significant factors that did Graham in, since an advantage stressed in court about the Graham plow had not been urged on the Patent Office. The Supreme Court disregarded the advantage as being "an afterthought.")

Nevertheless, the judge who first tried the Crosby case, Walter Bruchhausen, did not seem to attach much importance to the shift. Instead, he reviewed the history of the invention, the success it had achieved, the numerous invitations Crosby had received to speak about the invention at professional societies, and the acclaim it had been given by people in the industry. He called the invitations and demonstrations "some evidence that Mr. Crosby had produced a system, not then obvious to them [electronics experts]." Then he detailed the advantages of Crosby's system, including its monophonic compatibility, its ability to equalize the noise output in both channels, and its "high fidelity with complete separation at all frequencies."

Concerning the prior art marshaled by GE, Judge Bruchhausen said that although it indicated knowledge of the elements combined in Crosby's system:

> Those so skilled were unable to combine the elements as did Crosby in his single electronic system. He was confronted with the same obstacles and problems of

> FM stereo as were others devoting attention to the art. However, his ingenuity and efforts solved these problems. . . . Some twenty years had then elapsed since Major Armstrong opened the field of FM broadcasting and no one in the art appeared capable of producing the system now under attack during those two decades.

The judge found that GE "did not evince an interest in FM stereo until 1957 and at that time their representatives attended meetings where Mr. Crosby spoke on his system, attempted to obtain copies of his talks and met with him to obtain the details of his system." Relying upon Judge Learned Hand's admonition that the key to invention is "the act of selecting" materials and assembling them in a serviceable manner, he concluded that Crosby's patent was valid.

When the decision upholding Crosby's patent was announced, the triumphant inventor and his friends "had a small celebration," as one participant put it. Now there was hope of getting back in business again and of seeing their stock recover its value. This last was no small matter to people who saw a chance of becoming millionaires. Crosby himself owned some 250,000 shares and also had made personal loans to the company in its final days, which he hoped to see repaid. Furthermore, he had a contract with the company that called for him to receive 15 percent of the first $100,000 proceeds from the patent and 25 percent thereafter.

All these hopes were dashed when GE appealed. In his thorough eighty-eight page brief to the Court of Appeals for the Second Circuit, GE's attorney placed considerable emphasis on the "shift of position" by Crosby. He wrote that as a consequence of learning of the Boelens patent, Crosby "began the course of salvage which resulted in the claims relied upon here." Then he used the A&P language on several pages to attack the Crosby invention. One heading was "2 Plus 2 Equals 4"—a shorthand way of making the point that the invention did not exceed the sum of its parts under the A&P

ruling. Crosby was no Bert Adams, he insisted. All Crosby had done "was to put two known things together to get a more refined assembly having the sum of the known virtues of each." Consequently, "the Crosby system . . . falls into the pattern of classically unpatentable subject matter." And "there is not a single factor indicative of any unobviousness, or of any problem, or of any giant stride. . . ."

As for the acclaim Crosby had received for his speeches and demonstrations, the GE attorney said that the Graham decision, if it had done nothing else, had put an end to deciding the "unobviousness" question on the basis of "hoopla."

The answering brief by Crosby's attorney emphasized the performance of the Crosby invention itself. Paragraphs were devoted to "monophonic compatibility," "balanced noise," "high fidelity with separation at all frequencies," and "bandwidth." Only about one page was devoted to Crosby's "shift."

In a particularly forceful passage, Crosby's attorney said:

> When a problem has not yet been solved, the teachings addressed to that problem are voluminous, contradictory, obscure, and of little help. After the problem has been solved, it can be made to appear that some few obscure prior teachings could have been selected and could have been so interpreted and could have been so combined as to provide the solution.

The Court of Appeals made its decision on March 14, 1968, and referred in one way or another to the "shift." It decided that "Crosby achieved this compatibility by the use of a sum-and-difference matrixing, *which he believed he had invented.*" (italics mine)

As for the full separation of the two channels, which was emphasized in Crosby's brief, the Court cited Graham in support of a declaration that: "The belated emphasis on this feature . . . indicates that it is merely an afterthought."

Graham was also cited as a basis for downgrading the trial judge's conclusion that Crosby's results filled a real need in the industry, that he was acclaimed for this invention, and that the invention enjoyed commercial success. These, said the Court, were "not to carry too heavy a burden." The lower court was reversed; the patent was invalidated.

Crosby's attorney then applied for a hearing in the U.S. Supreme Court, making these arguments, among others:

> It is patent protection that makes the crucial difference to small organizations, such as Crosby Laboratories. It is patent protection that makes it possible for these small organizations to develop and grow in the face of the massive resources of the larger competitors. The limited monopoly of the patent is a . . . partial offset to a competitor's financial resources.

The request for hearing in the Supreme Court was, as is almost always the case, denied. Thus, on October 14, 1968, just a few weeks after Crosby's sixty-fifth birthday, the end came for his company. It was not the end for Murray Crosby, however. Leonard Feldman, who became associated with Crosby in 1959, remembers that on the eve of the FCC decision he teasingly asked Crosby what he would do if his system was turned down. "I'll tell you one thing I won't do," Crosby replied, "I won't do what the Major did." Crosby was referring to Armstrong's thirteen-story plunge to death one wintry night in 1954, following five years of fierce litigation over his FM patents—litigation that left him frustrated and bitter, his marriage a shambles, his financial resources threatened, and his enormous energy gone.

Murray Crosby retrenched and ultimately returned to the basement laboratory at his home. There, he continued inventing until his death on June 8, 1974. His final patent, issued shortly before he died, was for an ingenious digital system that, by detecting a subaudible coded signal, could keep track of radio and television commercials and even determine whether a song being played came from a pirated tape. To Feldman, this final invention of Crosby's illustrates the remarkable talent of a man who had matured in a pre-World War II engineering world and yet was mentally agile enough to make the transition to a new post-war digital technology. "Murray developed that system entirely by himself, just learning the whole digital technology from scratch and becoming an expert in it in his last years."

In November 1978, the Radio Club of America awarded its Armstrong Medal posthumously to Crosby for his pioneering work. Today that medal hangs in Mrs. Crosby's home near a large formal photograph of Crosby at a 1940 banquet in New York City's Waldorf Astoria Hotel. The banquet was sponsored by the National Association of Manufacturers to commemorate the 150th anniversary of George Washington's signing of the bill that laid the foundations of the American patent system—that very same system on which Murray Crosby had bet the last two decades of his life.

SIX

A Big Winner

Creativity is so delicate a flower that praise tends to make it bloom, while discouragement often nips it in the bud.

ALEX F. OSBORN

If there were a grand prize for winning patent validity contests, it would go to Bert Adams and his attorney for pulling off the remarkable coup of winning in the Supreme Court of the United States. There are, of course, numerous other patent holders who have laid their patents on the line, won in the lower courts, and then remained winners without ever having their patents subjected to Supreme Court scrutiny. One of these successful companies is the Ransburg Corporation. Not only is its story in the best Horatio Alger tradition, but it also illustrates how vital patents can be in building new companies in the face of competition from large, long-entrenched ones.

Like every youngster who ever took high school physics, Harold Ransburg learned that when he stroked a rubber rod with fur and touched it to a tiny, suspended pith ball, the ball jumped away. Then, when he rubbed a glass rod with a piece of silk and brought it close to the ball, the ball was attracted to the rod. While this hoary experiment must have intrigued hundreds of thousands of youngsters in its time, only a handful ever seized upon the scientific phenomenon to create a business. Chester Carlson was one, founding the billion-dollar industry of xerography upon the propensity of "like" electrical charges to repel one another and "unlike" charges to attract. Frederick G. Cottrell, an eminent scientist who in 1912 developed the electrostatic precipitator for use in industrial smokestacks, was another. And Harold Ransburg was a third. With electrostatics he revolutionized spray painting and, in the process, turned his father's small kitchenware company into a major producer of specialty industrial equipment, with sales of over $144 million for 1981.

Ransburg Corporation headquarters is a dun-colored brick building sprawling so low against the Indiana horizon that only a silver-gray water tower reveals its presence behind one of the few cornfields to have resisted the encroaching shopping centers and adults-only housing developments north of Indianapolis. Harold Ransburg, who retired as board chairman in 1976, came to work for what was then the Harper J. Ransburg Company in 1931. During that Depression year, he spray-painted cookie jars from seven o'clock in the evening until seven o'clock the next morning in his father's "ardent, enthusiastic sweatshop." By 1933, a year in which the firm shipped only $33,000 worth of goods and posted a $5,000 loss, Harold was in charge of painting the cookie jars, a dirty job in which most of the sprayed paint was wasted. Chagrined to see so much paint go out the exhaust duct, Harold resolved to find a better way to conduct the operation.

During the next few years, the young man experimented with various schemes for conserving paint. But nothing worked. Then in 1936—the year his impending marriage earned him his first paycheck ($30 a week)—his older brother suggested he see a friend who taught chemistry at a local extension college. "I guess my brother got tired of hearing me worry about what a tragedy it was to buy paint and then throw so much of it away," Ransburg says.

The friend referred Harold to his assistant, Harry Green, and after some discussion of the problem, Green asked whether Harold had ever tried electrostatic precipitation. That immediately rang a bell with Ransburg, for whom physics had been a favorite subject in high school, because he had recently visited a nearby Westinghouse Electric plant and had observed a demonstration of how smoke particles were precipitated onto a metal plate after being ionized in a high-voltage electric field (the Cottrell process).

Convinced they had found the solution, Harold Ransburg and Harry Green proposed electrostatics to Harper Ransburg. But they met with little encouragement. Ten years later the senior Ransburg told an interviewer, "Harold talked to me about it repeatedly, pointing out the extravagance of our system. That was nothing new to me, of course. I realized it, but was thoroughly convinced that if there was a better way, someone else would have thought of it. However, he [Harold] was so persistent I finally agreed to indulge his fancy to a limited extent just to keep him happy."

Indulging Harold's fancy meant allowing him $100 for materials and letting him and Green come in on Sundays to conduct experiments in a small room behind Harper Ransburg's office. ("Saturday was a regular working day if your name was Ransburg," Harold explains.) To conserve coal and to save the time they would have lost stoking the furnace, they left the building unheated. To conserve cash, they purchased an antiquated x-ray machine for $35 and scavenged its high-voltage supply, an enormous contraption of steel, porcelain, and insulating oil that weighed several hundred pounds.

Their idea was brilliantly simple: put an electric charge on the paint as it was sprayed from a conventional air-operated gun, and electrically ground the object to be painted. The paint, consequently, would behave just like the smoke in the electrostatic precipitator Harold had seen—it would be pulled toward the target rather than sprayed past it and wasted. Execution was not so simple, however. They had to figure out the proper way to put an effective charge onto the individual spray particles. They quickly learned that just connecting the spray gun to the high-voltage source was not effective.

It took several months of cold Sundays during the winter of 1937–1938 for them to discover that using as low an air pressure as possible and a charge of 100,000 volts permitted them to spray-coat all sides of a grounded metal salt shaker from a distance of ten or twelve inches. On March 6, 1938, when Ransburg's wife arrived at 5:30 P.M. to drive them home, they were too excited to leave and insisted she stay to witness their initial demonstration. First they turned on the gun for a few seconds and demonstrated that only the side of the shaker facing the gun became coated with paint. Then they switched on their x-ray power supply in order to charge four wires that were spaced evenly around the grounded shaker. Now when the gun sprayed, paint adhered to all the sides as well as to the top and bottom. When they left for home after three hours, they all knew that electrostatic paint spraying could work.

Like so many other inventors, however, they had little idea of how long and frustrating the journey would be from the first demonstration of their invention to its profitable commercial application. Another decade passed before electrostatic painting made its first dollar for the little company. It was not until 1951, after a major improvement on the basic invention had been introduced, that electrostatic painting really started to pay off.

Immediately after their original tests, Harold Ransburg and Harry Green set up a conveyor system for painting flower pots. They estimated that, even in the early, crude stage of the process, they saved one-fourth to one-third of the paint normally used. Representatives from big companies such as Chrysler and Du Pont came to observe what they were doing, but none could be sold on manufacturing the improved spray gun that Ransburg visualized or on using the process in their own plants. "People were afraid, and I think rightly so, that putting one hundred thousand volts in a paint spraying booth was very, very foolish, if not disastrous. We might burn the plant down," Ransburg muses.

Eventually, a discouraged Harry Green left. By the spring of 1942, Harold himself was ready to give up on electrostatic spraying for the duration of the war. With the United States having just surrendered Corregidor to the Japanese, mobilization was in full swing, materials were scarce, and nobody seemed terribly concerned about saving a few dollars on paint. Just then, however, a Du Pont executive telephoned to introduce a customer who wanted to see whether electrostatic forces could remove the little "tears" that formed when steel cartridge cases were dipped in and then pulled out of a bath of varnish. The test was successful. Ransburg ultimately leased to customers twenty-four electrostatic detearing units as well as some thirty-odd spray units for coating bazookas, ammunition cases and other war items.

Little money was made from these activities, but that didn't matter. Far more important was the fact that they permitted the company to receive an allotment from the War Department of scarce copper with which to build the high-voltage equipment. Work on the process could thus continue. This work eventually resulted in the accidental discovery that made electrostatic spraying a highly profitable business. The breakthrough was called the "Number 2 Process."

As Harold Ransburg recounts the story, an engineer was painting one of the high-voltage wire frames with an ordinary paintbrush. The voltage had been accidentally turned on, and when he moved the brush to within a few inches of the high-voltage frame, he was startled to see a spray of atomized paint particles being drawn off the end of the brush and deposited onto the frame. Harold's brother Edwin was called over to observe this entertaining bit of parlor magic. "That quickly rang a bell with Ed, who said, 'that's the way to paint things—no air gun to fool around with, just atomize the paint electrostatically.'"

They set to work, and in 1944 Ed Ransburg and William "Al" Starkey (the engineer with the lucky paintbrush) applied for patents on a trough-shaped device suspended over a conveyor belt carrying the objects to be spray-coated (the "workpieces" as factory jargon tags them) and on the corresponding method. Protruding through holes in the bottom of the trough were numerous needlelike wires that gave the assembly the appearance of some surrealist's sculpted paintbrush. The holes permitted paint in the trough to ooze down along the wires.

When the wires were electrically charged, paint particles were pulled off the tips of the wires and deposited on the grounded workpiece. By eliminating the compressed air, which in Harold's earlier process could blast as much as one-third of the paint away from the workpiece despite the presence of the electrostatic field, this process used virtually all the paint. As the method was perfected in the following years, Ransburg engineers estimated that 95 percent or more of the paint was actually delivered to the workpiece. This, in turn, permitted workers to spray in the open, without the need for expensive equipment to exhaust the paint fumes and to recover paint that missed its target.

The July 7, 1945 issue of *Collier's* magazine took notice of the work at Ransburg in its "Your Life Tomorrow" department. Along with an invention for replenishing the fizzy soda in a drink and another for a means of cooling factories by spraying water on the roofs, readers learned of the "electric paintbrush" that made use of "the same principles that drive tiny particles of electricity from negative to positive in radio tubes." It is likely, predicted the writer, "that entirely automatic painting machines will be designed to take automobile bodies, washing machines and almost any other irregular metal products, coat them electrostatically at high speed, dry them with infra-red lamps and then roll them out ready for assembly in a fraction of the time required to do the same job by hand."

Although that is a pretty fair description of how things are done today, in 1945 there was a long way to go before the vision was realized. The original Starkey-Ransburg apparatus had a number of weaknesses. It tended to clog and it did not produce an even coat because the droplets were formed at the tips of the individual wires. During the next six years, a small research department was built up at what became, in 1948, Ransburg Electro-Coating Corp., and a determined effort to improve the Number 2 Process paid off.

Eventually, Ransburg researchers—principally Edwin Ransburg and William W. Crouse—developed a practical, commercial device (and method) that produced an even coat without clogging. There was no clogging because there were no holes. Instead of being fed through holes and along wires, paint went to the hub of a spinning disc. Centrifugal force then took over, sweeping the paint along the surface of the disc to the outer edge. There, because of the electric field, the paint was atomized and pulled from the edge for application to the workpiece. The paint particles flew off from almost the entire edge, in a spray much finer and denser than was possible using individual wires. Thus, a much more uniform coating was produced.

By 1956, Ransburg Electro-Coating Corp. was on its way toward becoming "No. 1 in the electrostatic finishing industry." In an enthusiastic review of Ransburg's history, written in 1964, *Indiana Business and Industry* attributed the company's preeminence to "the sheer weight of its engineering excellence" and its willingness to battle in the courts to protect its valuable patents.

Starting in 1956, Ransburg battled infringers in just about

every industrialized country of the world. Amazingly, the little company won every important contest. In the United States alone, some ten infringement suits went through trial to decision. Perhaps another one hundred cases were settled when the company threatened to sue or before a decision could be reached in court. A major reason for the numerous legal battles was that Ransburg's electrostatic processes were so inherently simple that they could easily be duplicated. According to Merrill Johnson, the firm's former general counsel (and later board chairman), hundreds of companies—large and small—tried to do so over the years.

"In Japan we had over four hundred infringers, and at one time there were fourteen lawsuits brought against us by seven or eight blue-ribbon Japanese manufacturers who had banded together to try to break our No. 2 Process patents." But after the Japanese patent court ruled in favor of Ransburg, declaring its patents valid, "it became impolite not to settle, and the smaller infringers literally lined up in a queue to pay us the computed damages."

Johnson estimates the Ransburg Corporation collected some $2 million in damages in Japan, while in the United States the total probably came to over $4 million. Litigation elsewhere—in Austria, Belgium, Canada, Denmark, France, Germany, Great Britain, Hong Kong, Italy, Norway, Sweden, South Africa, and Switzerland—very likely brought in another $3 million. In sum, "We never lost an infringement suit on the basic No. 2 Process patents," he says proudly.

Johnson relishes telling how big German and Japanese companies could not believe "some little pipsqueak company from Indiana" would travel all the way over there and go through the expense of suing them. Their idea of a fair settlement, he says scornfully, was ten cents on the dollar. Ransburg would have none of that. One case against a giant German manufacturer dragged on so long that the two sides eventually agreed to arbitrate. A panel of three retired judges from the German supreme court ruled in favor of Ransburg.

Then, however, the German company changed its equipment and claimed it was not infringing. "So we sued for criminal breach of the agreement to arbitrate, and when the German court agreed, boy, did that shake them up! Their board of directors took the matter away from the patent people, who thought they had been playing it pretty smart, and negotiated a realistic settlement with us. They didn't want it public knowledge that they had dealt in bad faith."

Not surprisingly, the hundreds of thousands of dollars and countless man-hours invested in this legal marathon were so great that there were weeks when Ransburg executives wondered whether they were in the business of lawsuits rather than the manufacture of spray-painting gear. At times, all four attorneys in the company's legal department would be working on a case, together with outside counsel.

Emery Miller, an ex-Purdue University Ph.D. physicist who led the twenty-man research department, recalls that during the weeks preceding a trial, and while the trial was actually in progress, his time as well as that of perhaps one-fourth of his technical people would be devoted almost exclusively to helping prepare testimony and to conducting experiments, running tests, and setting up equipment for the elaborate courtroom demonstrations that were so important. "It was something we had to do," he reflects. "You either did that or went out of business."

All in all, Ransburg's patents were among the most extensively litigated in history, ranking almost in the same league as the Bell telephone patents, which between 1879 and 1896 were tested in hundreds of cases.

Johnson is convinced that had it not been for its electrostatic spray-coating patents and the upholding of these patents in court, the Ransburg Corporation would not now exist. Moreover, he believes that even if these patents had been upheld, Ransburg today might well have been just another tiny company struggling for a meager share of the spray-painting market. What made the difference, he maintains, was the company's carefully thought-out licensing policy, under which Ransburg did not permit other paint-sprayer manufacturers to sell its automatic spray-coating systems in return for royalties based solely on the value of the equipment sold. This would have been the conventional thing to do. What Ransburg did instead was offer licenses to anyone—other spray-system manufacturers as well as users of the equipment—in return for *continuing* royalties based on the amount of paint saved. When a potential customer appeared, Ransburg engineers investigated his needs and then set up a pilot operation demonstrating how much paint he could expect to save. If the prospect liked what he saw, Ransburg licensed him to use the system, which it installed and helped maintain. The customer paid an initial fee, which varied with the equipment licensed, plus monthly fees representing a portion of the paint savings.

Such an unconventional policy did not endear Ransburg to either its competitors or to those industrial giants that were used to buying systems outright. Only a strong patent position kept Ransburg from having to knuckle under to demands that it follow a more standard policy.

The initial court battle to test this licensing policy, together with the patents themselves, was precipitated by Binks Manufacturing Company, one of the largest suppliers of paint-spraying equipment in the United States. Advertising "everything for spray painting," Binks in July 1956 stated that it was ready to market a paint sprayer that electrostatically atomized paint without a compressed air gun. Binks applied to Ransburg for a license, but when Ransburg sent its Number 2 Process license forms, Binks objected. Binks wanted to make only a single royalty payment, at the time when the equipment was sold, so that its customers would be free of future royalty payments.

Twenty days after Ransburg refused this request, Binks, believing itself "foreclosed from entering the market of airless electrostatic coating," sought a declaratory judgment that the

Starkey-Ransburg Number 2 Process patents were invalid. Misuse of the patents, violations of the antitrust laws, and unfair competition were also alleged. Ransburg fired back with a suit for patent infringement, and the battle was joined.

The trial took the better part of three weeks in the United States District Court for the Southern District of Indiana, Indianapolis Division. During the course of the trial, Ransburg's patents were bombarded with prior art going back as far as 1850, when an Englishman showed how electrostatic forces could be used to spin threads from molten sealing wax. More closely related was a 1934 patent by an electrical engineer named William Darrah. Darrah, in a deposition, said he had used his electrostatic system to spray colored water on paper more than thirty years earlier. His testimony might well have been the end of the Ransburg patents, except that Darrah was then asked what he had done with his invention after that demonstration. He replied that he had not been able to think of any application, and consequently, "I just let it rest."*

On June 16, 1959, Judge Cale J. Holder decided for Ransburg in all respects. The Ransburg invention, he ruled, was "a broad, basic invention fundamentally different from any previous spray-coating process of record here." Judge Holder found nothing illegal about Ransburg's licensing practices and decided that the company dominated the market for airless electrostatic spray-coating systems only "by virtue of the monopoly created by its patents."

All told, Judge Holder approved ninety-four detailed findings of fact and eleven conclusions of law. Since Binks believed many of these were erroneous, it appealed. Alleging

---

*Besides Darrah, the patent that might have come closest to tripping Ransburg up was a Western Electric Company patent on the basic idea of putting an electric charge on paint particles. Discovered by Ransburg when it was cited by the Patent Office against the application covering the original process, the patent was never commercialized by the giant electrical manufacturer. Although Ransburg engineers believed it was an impractical device, the company bought it from Western Electric for a few thousand dollars in order to avoid any future claims of infringement.

that the patents were invalid for lack of invention over the prior art, Binks cited Ransburg's own detearing process, marketed during World War II. The implication was that once one learned to remove tears of paint electrostatically, it was obvious that paint could be sprayed electrostatically. Ransburg replied that the Starkey-Ransburg invention had evolved several years later, and then only as a result of the fortuitous paintbrush incident. In an eloquent reminder of the snares of hindsight, Ransburg's attorney wrote:

> The fact that experienced personnel working with electrostatics to improve spray-coating techniques had the detearing process in front of them for several years without receiving from it any hint as to how spray-coating could be improved is compelling evidence that invention was present in what Starkey and Ransburg did. Ex post facto wisdom may make the step . . . to Starkey-Ransburg look short; but ex post facto wisdom, as the courts have repeatedly held, is not to be used in measuring invention.

Except for the minor detail of reversing the district court's allowance to Ransburg of attorney fees, the appellate court supported Ransburg completely. "None of the prior art teaches nor suggests the combined atomization, dispersion and deposition of coating material through utilization of electrostatic force," the court ruled. It found that the patents combined "old elements in a novel manner productive of new and useful results." It decided that the district court had applied "the correct criteria of invention," and it cited the A&P decision in support of this statement!

The Number 2 Process patents thus survived their first and most crucial challenge. Ransburg was now poised to pile victory upon legal victory. Among the giants going down to defeat were Ford and General Motors. Unhappy with Ransburg's licensing policy, according to Merrill Johnson, both spent several years trying to produce electrostatic spray-painting systems that would not infringe on the small company's patents.

After its own development effort had borne little fruit, Ford enlisted the University of Michigan's prestigious Engineering Research Institute. The institute worked with Ford from 1955 through 1958 before abandoning the effort to produce a high-pressure paint-atomizing system for electrostatic finishing. The auto company then went ahead with a system that, in 1965, was found to infringe the Ransburg Number 2 Process patents. Shortly after Ransburg won this suit against Ford, General Motors settled out of court for over $2 million. GM and Ford subsequently became two of Ransburg's best customers.

In 1974 fees from licenses reached a peak of nearly $11 million. Because related costs were minimal, the company claimed to be one of the most profitable in the United States (in terms of return on stockholder's equity) between 1972 and 1976, when the Number 2 Process patents began expiring. By 1981, when its Japanese patents ran out, Ransburg had garnered about $200 million in royalties since 1951. Not bad for a company whose annual sales didn't break $30 million until 1972. But it was only by the readiness to sue each time some firm balked at the terms of the proffered license that Ransburg was able to protect these royalties, for it could not let one company off and expect others to continue paying.

Why was Ransburg so successful in court? In the first place, the crucial early cases came in the late 1950s and early 1960s—after the 1952 Patent Act but before the Graham decision made life far tougher for patent owners. In the second place, much of the litigation took place in the Seventh Circuit, which in those days was one of the most pro-patent circuits in the nation. (In recent years, it has held considerably fewer patents valid.) Third, the Number 2 Process is dramatic to watch. Workpieces, such as unpainted hose nozzles, travel via an overhead cable through the painting area, where the dull gray metal instantly turns a beautiful smooth yellow. Only by peering closely does one see that paint is being sprayed, and no residue is left on the floor or the walls. Everything happens so quickly that there is a magical quality to it.

Unfortunately, judges today rarely have the time to visit an inventor's laboratory or factory to see a demonstration like this. In Ransburg's case, however, several did. In one trial, in 1962, the judge spent an astounding three days out of the courtroom viewing demonstrations of Ransburg's equipment as well as some of the alleged prior art. During a 1967 trial in the British House of Lords (which also constitutes Britain's highest court), the five judges drove from Parliament under police escort to Ransburg's laboratories. Without the striking exhibition that followed, they might have had a more difficult time deciding that the patents were valid and had been infringed.

Finally, Ransburg had excellent counsel, as well as sheer good luck—which has won more legal cases than one might think. Had William Darrah, for instance, not answered as he did, Ransburg could well have lost the crucial Binks case.

I cannot end the tale of what is now an expanding Ransburg Corporation without harking back to Harold Ransburg, the young man who, without any formal technical training, was convinced that he had a great idea and—even more important—was willing to spend his entire waking hours thinking about it. "Never for a minute did he doubt the capability of its operation, and anybody who wouldn't buy it was out of his mind as far as Harold was concerned," says Emery Miller. Indeed it was likely that had one of the companies to whom Ransburg first showed his invention offered $100,000 for it, Harold's father would have grabbed it. Years later, of course, Harold Ransburg loved to repeat his dad's answer to people who asked how his son came up with electrostatic spraying: "The smart boys knew it couldn't be done, but Harold didn't know any better, so he just went ahead and did it."

The Ransburg Corporation is today diversifying into new

fields, principally the manufacture of industrial robots. This is a mammoth undertaking requiring vast amounts of invention, engineering, and capital. To be preeminent in the field, in which Japanese companies are researching and spending heavily, and which may (Ransburg Corporation believes) have a world-wide $20 billion market by the year 2000, Ransburg in 1980 formed with Renault a joint venture that Ransburg controls. The result is a U.S.-based robot company called Cybotech.

In Ransburg's drive for new products, it spends at least 4.5 percent of its sales—a greater proportion than most manufacturing companies—on research and development.

It was endeavors and expenditures such as these that I believe the writers of the U.S. Constitution hoped to stimulate when they authorized Congress to "promote the progress of . . . useful arts. . . ."

SEVEN

The Road Toward Confusion

. . . the value of it is greatly enhanced by the simplicity of the works, which, when seen and explained, may be executed by the most common mechanic.

GEORGE WASHINGTON

For several years after the 1966 Graham opinion, with its famous preface interpreting the obviousness section of the 1952 Patent Act, many patent experts believed that the strange and confusing doctrines of the A&P case were dead. They felt that some stability had been achieved (although with "strictness" and "secondary" factors) as to what was and was not patentable. Still, there lurked in the Graham opinion a possible ambiguity about whether *any* judge-made tests should be used. Moreover, the opinion had not actually driven a stake through the A&P case—had not declared *explicitly* that A&P was wiped out.

The door was therefore left ajar for A&P concepts to enter once more and plague inventors as they had in the past.

When such a door is left open, it is likely that infringers will walk through. Actually, they rode through in style on a road unknowingly built by an inventor whose main goal in life was to pave asphalt highways. His road-paver case reached the Supreme Court in 1969, and in its opinion the Court (while speaking also of unobviousness) went back to, and specifically cited, the A&P concepts. The result was that a strong, albeit confusing, antipatent argument, drawing upon both Graham and A&P, could thereafter be used by defense lawyers.

Charle (pronounced "Charlie") Neville was the kind of fellow people remember affectionately as a "character." Neville's patent attorney, Walter Blenko, Jr., remembers, "With Charle nothing was ever quite the way it seemed. His appearance was a little different, his manner of speech was a little different. He had an inquisitive mind and he was never inhibited by the ordinary ways of thinking. But then, if you look upon invention as doing something that's not obvious to a person of ordinary skills, it simply has to be an act of someone who is not behaving in a logical, straightforward manner. What he does only seems logical after he's done it."

Sometime prior to 1954, the six-foot-plus, 250-pound Neville huffed his way up five flights of stairs to see an engineer in the Pentagon office of "Chief, Pavements and Railroads Section, Director of Installations, Headquarters, United States Air Force." Charle had apparently overlooked the escalators, an omission typical of the burly inventor, who was often in too much of a hurry churning out new ideas to pay attention to the lesser details of life.

On this particular day, Neville was anxious to explain to the engineer, who was in charge of the maintenance and construction of Air Force bases, how he could repair airfield runways better by using a certain kind of radiant heater. Neville had developed a machine to solve a problem that had long plagued roadbuilders: how to get a lane or strip of hot new asphalt, or "blacktop," to bond properly with the edge of a newly laid lane that had already cooled. The difficulty arose because paving machines were not as wide as the road; consequently, unless two pavers were operated it was practically impossible to lay the second lane while the first was still hot. The result was the creation of a "cold joint" where the two lanes touched, leading eventually to cracking and deterioration of the pavement when water penetrated the cracks.

Other devices had been patented for this purpose, but, as a court of appeals later observed, they did not work in commercial operation. Consequently, before Neville came on the scene, the practice was to lay down one strip of asphalt, then cut back the edge with pneumatic hammers or other devices, paint the fresh edge with hot asphalt, and lay down a second strip next to the first one. This scheme was expensive and "not a solution of the cold-joint problem, but the best known method of minimizing its adverse consequences." Neville, the court said, turned away from current concepts in his search for a solution and went back to "the discarded notion of preheating the old material."

Specifically, Neville selected a particular type of radiant heater and used it in conjunction with a road-paving machine so that two contiguous lanes could be laid down and the proper joint obtained between them. This was accomplished by laying one lane, then returning to the starting point and effectively reheating the edge of the first lane while the second one was being deposited. In this way, both lane edges would be at approximately the same temperature, thereby ensuring good bonding.

Years later, in conjunction with the court fight over Neville's patent, the Air Force engineer explained that he did nothing about Neville's equipment because he did not believe it would do the job. He later changed his mind after encountering a California public works official who informed him of a successful demonstration that Neville had arranged in that state. The state official also had not believed Neville's machine would work until he saw it for himself.

The road that brought Charle Neville to inventing and to the Pentagon had been a circuitous one. A barber's son, born in Chatham, Ontario, in 1895, Neville's formal education ended in grammar school. By the time he was thirty he had prospected for oil, driven a taxi, taught mechanics, managed a truck engine manufacturing company, and started his own trucking business. When that failed, he went to work for a California insurance company that wrote performance bonds for highway contractors. There the gregarious entrepreneur was drawn into extracurricular forays in the contracting business, helping contractors locate raw materials and prepare bids. One day, after he and a friend had bid successfully on a highway construction job near Santa Barbara, the friend pulled out and Charle was left holding the grading subcontract. Always the optimist, and undaunted by his lack of machinery, he rented four-horse teams from local farmers to move the dirt, scrounged whatever light equipment was needed, and went into the road-grading business.

Neville divided his energies between building roads and writing bonds for several years, during which time he and his wife separated. His son Stewart recalls that he and his brother "scrambled to keep things together" and that he didn't see his father for a number of years. Their extraordinary reunion came about after Stewart had completed a World War II stint with the Army Corps of Engineers and had gone to work for a builder of prefabricated homes in Sacramento. "One day I was driving to work along Highway 40, which used to be the main road to Sacramento, when just before crossing the Sacramento River I spied my father on the side of the road. So I pulled over and we spent the afternoon together."

The account Charle Neville gave his son that long-ago summer afternoon was the story of how he started his career as an inventor. To appreciate this story, it is necessary to take a quick detour through some elementary aspects of road building. A so-called asphalt (blacktop) road actually consists of stone, sand, and dust (the aggregate) held together by 5–10 percent asphalt, which is essentially a petroleum-based binder. When Charle Neville built roads, he rented road mixer machines to combine hot asphalt with the aggregate and deposit the mixture on the roadway. The mixture was then spread, "laid to grade," and rolled flat.

One day Neville found himself with aggregate that would not mix properly. Apparently the wrong type of asphalt for that particular mixing operation had been delivered. His supplier, Shell Oil Company, offered to resupply him with the correct type, but Charle was in a hurry to lay a two-mile stretch of road. He urged Shell to send him a large batch of an asphalt softening compound that the company had been trying to market for some time with little success. The softener dissolved old, dried-out asphalt so that it could be reworked. An amateur chemist, Neville had been experimenting with the Shell compound in his garage and was convinced that, by applying precisely the right amount of softener, he could use it to "reclaim" cracked, worn-out asphalt roads. Guaranteeing to replace the entire two miles if the attempt failed, Neville obtained the state's permission to do the stretch of road his way.

When the experiment succeeded, Neville received an exclusive license to sell the product, and formed a company called Reclaimix. Next he obtained $25,000 from the State of California for a formal trial of his process on a deteriorated road near Sacramento. A photo taken in the summer of 1950 shows Charle, Stetson hat in hand, with his arms folded confidently on top of a road divider bearing a large sign:

THE SURFACE OF THIS
HIGHWAY IS BEING
RECLAIMED AND REMIXED
BY
THE RECLAIMIX PROCESS
CHARLE NEVILLE
ASPHALT SOFTENER
SUPPLIED BY
SHELL OIL COMPANY

The results of Neville's labor were reported by a California state highway engineer in an engineering magazine later that year: "The experimental section has been under traffic for two months carrying an average of 17,000 vehicles per day and is in excellent condition. . . . In all physical respects the work has been highly successful."

In December 1950, Charle applied for a patent on his "Process for Reclaiming Asphaltic Pavement and a Composition So Formed," which was granted five years later. A flurry of newspaper publicity ensued. "Shell announces new 'asphaltic softener' process [to] cut highway repair bills up to 50%," wrote the *Sacramento Union,* adding that the process had been developed jointly by Shell and Reclaimix Inc. An article in the *Los Angeles Daily News* called attention to the "monstrous looking machine which rolls along eating up a whole lane of ripped up road and leaving a rejuvenated road behind it."

The *Daily News* was actually describing the associated crushing machine, but it had the concept right. The perfecting of a single machine that would "eat" old, cracked roads and regurgitate new ones was a dream Charle Neville would pursue for the rest of his life. Heretofore, highway engineers had been forced to choose between resurfacing the old pavement—a makeshift solution that was not very satisfactory—or tearing up the road, carting it away, and building a brand-new highway in its place.

Stewart Neville joined his father in Reclaimix. But although they were able to contract for several jobs and provide substantial savings to some customers, they were unable to turn the scheme into a commercial proposition. Consequently, they decided to take a different tack and sell franchises for the Reclaimix fluid. It was this decision that brought Charle together with his entrepreneurial partner, J. Bennett Walker.

Except that Ben Walker came from a well-to-do Pittsburgh family and was about ten years younger than

Charle Neville, the two men were much alike. Somewhat on the portly side, though not as tall as Neville (and far more dashingly dressed), Walker was equally opinionated, also a great optimist, and possessed a good sense of humor. A "marvelous" salesman for his father's Pittsburgh Water Heater Company, he subsequently built homes, tried a frozen food venture, and became a manufacturer's representative selling water heaters and large infrared space heaters.

Sometime around 1954, Walker's brother spotted Reclaimix's franchise offer in *The Wall Street Journal.* He responded for Ben, who, after being advised by Pittsburgh's Director of Public Works that "it would be a good thing if it was possible to do," seems to have been the only person to maintain a serious interest in Neville's product. (Pittsburgh's hilly terrain and bitter winters made it something of the world's pothole capital.) Packing a can of fluid, Stewart hopped a plane to Pittsburgh, where he successfully demonstrated the softener for Walker on a pile of asphalt in one of the city's garages. An opportunity to fix a Pittsburgh street followed. This led the Nevilles and their new partner into a profitable business selling their softener by the tankcar load to the Allegheny Asphalt Paving Company. Allegheny had piles of used asphalt that it had taken off old streets and for which it had had little use—until Charle came along. Under his guidance Allegheny built a plant to reclaim the asphalt with the Reclaimix fluid.

Walter Blenko (Neville's patent attorney) began working with Charle to write the patent on his road paver in 1957. He remembers how he would prepare one draft, only to have Charle, who was always out in the field experimenting and refining his equipment, return with new ideas that would lead to yet another draft. "You never knew where a discussion with Charle Neville would lead," recalls Blenko. "Just about the time you thought you had everything straightened out, he'd be back with some more ideas."

The patent application they finally filed, in early 1959,

described two different machines. One was intended to realize Neville's dream of a single machine that first softened the old road with radiant heat, then treated it with the reconditioner, and finally mechanically worked, leveled, smoothed, and compacted it. The result: a seemingly brand-new highway.

The second machine, the one Neville tried to sell the Air Force on, was not for old roads but for building new ones; it was this machine that became the subject of the court fight.

To exploit both of these inventions, the Nevilles and Walker incorporated Pavement Salvage in 1958. They were fairly successful selling a certain radiant heater for use on asphalt-paving machines to solve the cold-joint problem. But they were much less fortunate with their old-road reclaiming machine, of which there were several versions. One was towed by a jeep. Another was a mammoth self-propelled vehicle that people sometimes described as a "Rube Goldberg" contraption. Its thirty-foot-long body rode about four feet above the ground. Slung underneath were first, a bank of fifty-six radiant heaters, then the gear for applying the reconditioner and raking the surface of the road, and, finally, a row of heavy-duty tires that served to compact the surface.

A third version, almost as long as the second, was built around a tractor with a stretched frame. Underneath the elongated carriage hung the heaters, which were followed by a pawlike rotary-blade assembly that could be lowered to the street level so as to agitate the softened asphalt while the reconditioner was being added.

The machines were plagued with the kinds of bugs that bedevil most prototypes. Once, wind kept blowing out the heater flame, causing the company to fall so far behind on the job that a lien was put on the machine. On other occasions, the heaters set fire to the reconditioner fluid. The most striking problem occurred when the machine ignited an Air Force base runway as it was being leveled. "That really flipped me," says Stewart, who explains that the fire was caused by a sealer that, unbeknownst to the equipment operators, had been applied to the runway surface earlier.

The little company was besieged by more than the technical difficulties that are almost inevitable with any invention. Before his patent was actually issued, in September 1962, Charle Neville died in his sleep. Eleven months later, Pavement Salvage initiated the first infringement proceedings. Not long afterward, Stewart left the company, disheartened, he says, over the company's prospects once litigation had commenced.

But feisty Ben Walker had no intention of giving up. He traveled abroad selling the heaters, and was determined to see the litigation through. Though discouraged by the mechanical difficulties, he kept on trying to find a big manufacturer willing to build the reclaiming machine.

The Road Paver in Court

One of the great myths of inventing is that it takes place through giant leaps. People believe that James Watt invented *the* steam engine, Thomas Edison *the* electric light, and Robert Fulton *the* steamboat. But, in reality, Watt invented an *improved* steam engine, using a separate condenser, while Edison invented a *practical* incandescent lamp using a high-resistance carbonized filament. As for Fulton, thirty-five steamboats were built before the *Clermont* sailed the Hudson. It was Fulton's "supreme luck," wrote sociologist S. C. Gilfillan, to do his work at a turning point in steamboat history, "but his work was no more significant technicly [sic], nor needed more genius, than the steamboat work of several men before him, and many after him."

It is taking nothing away from these and the hundreds of other pioneering inventors to bear in mind Isaac Newton's famous statement, "If I have been able to see farther than others, it was because I stood on the shoulders of giants." That declaration is as true in the world of invention as it is in science. As Supreme Court Justice Felix Frankfurter once put it, "Seldom indeed has a great discoverer or inventor wandered lonely as a cloud. Great inventions have always been parts of an evolution, the culmination at a particular moment of an antecedent process."

Charle Neville's inventing, like that of most inventors, proceeded in stages. An early step was taken when he conceived that a *certain kind* of propane-fired radiant heater could do a great job in one of the machines described in his patent, the one for renewing old blacktop roads. Later, Charle took another step by discovering that this same radiant heater could be used successfully to solve the cold joint problem in new blacktop roads.

The radiant heater Charle used had been invented several years earlier by a German named Gunther Schwank. It served many purposes, a number of which were listed in Schwank's patent, but prior to Charle's conception it had never been used to heat a highway. Instead, asphalt highways had commonly been heated by direct-flame burners that tended to harm them, or—much earlier—by radiant heaters that were grossly inefficient and had long since been abandoned.

When Neville's patent was finally issued on September 25, 1962, all its claims required radiant heaters of the general Schwank type, combined with other components of Charle's two machines. As is most always the case, these other components had long been known and were what inventors call "old in the art." In fact, it took over three years to get the patent because the examiner would not accept the initial claims, which called for radiant heaters of no particular genre. It was only when the claims were changed to specify Schwank-type heaters that they were accepted and the patent was issued.

In the interim, Pavement Salvage proceeded to sell Schwank-type heaters for combination with asphalt-paving machines. By April 1965, over one-half million dollars' worth had been disposed of in twenty-nine states and three foreign countries. In the early 1960s, however, an old-line heater manufacturer started to make and sell, for asphalt-paving machines and other purposes (such as patchers), heaters that Pavement Salvage believed infringed Charle's patent when contractors mounted them on their asphalt-paving machines. This manufacturer was Aeroil Products Company, Inc., a South Hackensack, New Jersey, firm. Among its customers was a paving contractor named Anderson's – Black Rock, Inc.

Pavement Salvage filed a patent infringement suit against Anderson's – Black Rock on August 28, 1963 in the U.S. District Court at Charleston, West Virginia.* After a two-day trial in April 1965, the judge ruled against Pavement Salvage. Charle Neville's patent claims for solving the cold-joint problem in new highways were declared invalid. (This, incidentally, was the only question the District Court—or any other court—ever decided. At no time did any court decide the question of whether or not there was infringement.)

Pavement Salvage thereupon took its case to the U.S. Court of Appeals for the Fourth Circuit. On November 16, 1968, that Court voted two-to-one to reverse the trial judge and hold Neville's patent valid. The majority opinion focused solely on the question of obviousness, and considered it noteworthy that earlier patented devices were not commercially feasible. Although Neville did not invent the radiant energy generator, said the Court, his proposal was sufficiently radical to be greeted with incredulity by the paving industry. After citing the Air Force engineer's expressions of disbelief, as well as testimony by a research director for the Pennsylvania Department of Highways, the Court said:

> The fact that experts in the field received Neville's disclosure with such skepticism and disbelief strongly indicates that, at the time, a person of ordinary skill in the field would not have sought the answer in the disproven method of heat treatment.

*When a customer of a manufacturer is sued for patent infringement, it is conventional for the manufacturer to defend the suit at its own expense and pay any damages. Aeroil did so here and was the real defendant.

Continuing its attack on the contention that the invention had been obvious, the Court of Appeals stated:

> There was nothing in the junk pile of prior art heat treatment patents to make it obvious to anyone that they supplied the ultimate solution. That this is so is forcefully demonstrated by the duration of the fruitless search, by the skepticism and incredulity with which experts in the field received Neville's disclosure, by its commercial success after demonstrations dissipated the disbelief of the experts. . . .

Then the Court of Appeals made a crucial statement about the trial judge's decision:

> The District Court was not unimpressed with all of this, but *it viewed the question as if it were known that the answer lay in heat treatment and that a generator of the type described by Neville would do the job.* Had it been known that the result could be achieved by a generator and a paving machine handled separately but cooperatively, combining the two in one machine for the sake of convenience would have been an obvious step, but *that was not known.* (italics mine)

Anderson's—Black Rock petitioned the U.S. Supreme Court, which agreed to hear the case during its October 1969 term. This was bad news for Pavement Salvage, since patent-validity cases in the Supreme Court were almost always lost by patent owners.

Anderson's—Black Rock, Inc. v. Pavement Salvage Co. (usually referred to as the Black Rock case) was the first patent-validity case heard by the Supreme Court in almost four years (since Graham, Calmar, and Adams). On December 8, 1969, the Supreme Court issued its unanimous decision (the Chief Justice not participating) invalidating the patent. The Court, in an opinion written by Justice William O. Douglas, said:

> There is uncontested evidence that *the presence of the radiant-heat burner in the same machine with the other elements is not critical or essential to the functioning of the radiant-heat burner in curing the problem of the cold joint.* For it appears that a radiant-heat burner operating in a tandem fashion would work as well. The combination of putting the burner together with the other elements in one machine, though perhaps a matter of great convenience, did not produce a 'new or different function' . . . *within the test of validity of combination patents.* (italics mine)

Because the Supreme Court never makes public the discussions that lead up to a decision, there is no way of reconstructing the reasoning that led to Justice Douglas' opinion about Charle's paving machine. One can only interpret the opinion itself. On this basis Justice Douglas appears to have used the same approach as the trial court; that is, treating the situation as if someone had already used the particular (Schwank-type) radiant-heat burner on asphalt pavement and that Neville had merely hung it on an existing paving machine. Justice Douglas disregarded the statement by the Court of Appeals that Charle was the first to use this particular burner on pavements in any manner, whether or not hung on the side of a paving machine.

As for the experts' testimony of being amazed and disbelieving at the success of Charle's machine, Justice Douglas said in a footnote that these statements focused exclusively on the properties of the radiant-heat burner and not on the combination of the burner with the other elements of the paver. This was the closest that the opinion came to discussing signposts.

In short, Justice Douglas was applying what he called (at

the end of the above-quoted passage) "the test of validity of combination patents." That test is—under A&P doctrines—more "exacting" than that applied to something (if it could even exist) that is not such a combination. In subjecting the patent to that test, Justice Douglas dismissed the kinds of signposts that Judge Learned Hand followed and that the Court of Appeals relied on when it held in favor of Pavement Salvage. Without any substantive discussion of these crucial determinants of obviousness, he simply concluded that "the combination was reasonably obvious to one with ordinary skill in the art." Wrote Justice Douglas: "It is, however, fervently argued that the combination filled a long felt want and had enjoyed commercial success. But those matters 'without invention will not make patentability.' A&P Tea Co. v. Supermarket Corp. . . ."

For people not connected with Pavement Salvage—for inventors everywhere—the most important thing about this case is not what the Supreme Court decided, or even whether the decision was correct or not. What is crucial is what the Court said—what legal arguments it gave defense lawyers.

What the Court said in this case revitalized the A&P concepts, *intermixing* those concepts with obviousness, and thus laying the foundation for a more devastating Supreme Court decision and opinion. In addition to the statement about the test of validity of combination patents, the Court declared:

> A combination of elements may result in an effect greater than the sum of the several effects taken separately. No such *synergistic* result is argued here. . . . We conclude that while the combination of old elements performed a useful function it added nothing to the nature and quality of the radiant burner already patented. (italics mine)

These concepts are straight out of A&P. There is but one new word—"synergistic." As the quote makes clear, that word was only used as a shorthand expression for A&P's

whole-must-exceed-sum-of-parts test. It should, therefore, have had little effect, but in the typical way that strange doctrines evolve in patent law—when it is not made absolutely clear that there is to be only a single, statutory, test of patentability—defense lawyers were able to use this unfortunate word to create massive confusion in years to come.

For almost seven years, many patent lawyers and informed inventors believed that the Black Rock revitalization of A&P concepts was an anomaly that would mean little. That dream, however, was to be destroyed in 1976 by a much more famous Supreme Court case that again relied strongly on A&P and again used the word "synergistic."

EIGHT

The Cow Barn Disaster

. . . I could have constructed a large steel grasshopper propelled by successive explosions of gunpowder, and ridden it into the blue realm of the stars.

ROSTAND'S CYRANO
ANTICIPATES APOLLO XII

April 20, 1976, was a miserable, rainy Tuesday in Tulsa, Oklahoma, and Donald Joseph Gribble was thoroughly soaked from the thunderstorms that had gusted around the "World's Oil Capital" all day. He was glad to be returning to his hotel from nearby Okmulgee, where he had participated in opening ceremonies for a beef feedlot facility his company had recently installed. Entering his room at the neo-Arthurian Camelot Inn on South Peoria Street and Interstate 44, he dropped into a chair and switched on the CBS Evening News. Walter Cronkite was announcing that Timothy Leary would be free in June and that seventy-six airline pilots had asked the FAA to ban smoking in the cockpit. Then came a lighter item:

> The Supreme Court has settled the question of just how original is the idea of a machine which dumps water on a dairy-barn floor to wash away cow manure. The Court said it's not original enough to merit a government patent. . . . Justice William Brennan, ruling against a firm called AGPRO, said exploitation of the principle of gravity adds nothing to the sum of useful knowledge. Next case.

The Supreme Court decision delighted newspaper headline writers, such as one for the *Los Angeles Times* who punned "Barn Cleaner Fails to Hold Water." But to inventors and patent attorneys — as well as to Joe Gribble — the decision was much less amusing. AGPRO was Joe Gribble's company. He and Henry Bennett had invented the barn-cleaning apparatus, and its installation in dairies and feedlots was a vital part of his small business.

One hundred and five miles northeast of Dallas, Route 82 bolts eastward from Bonham, Texas, past the beef ranches and neat farms around Honey Grove ("The Sweetest Town in Texas, pop. 1,853"), and on into the 25,000-population rail junction that bills itself "The Second Largest Paris in the World." The Texas Paris boasts thirty-five manufacturing companies, some nationally known. But tiny AGPRO would only be familiar to people who use the specialized dairy and farm equipment AGPRO manufactures in its two-story concrete block building six miles east of the town. Joe Gribble's small, spare, second-floor office overlooks the parking lot, a large metal grain silo, and 460 acres on which wheat, milo, and soybeans are grown — "strictly to help the profit-and-loss statement," says the compactly built fifty-year-old former high school wrestler.

Gribble's life has been tied to agriculture since his boyhood on a Molalla, Oregon, dairy farm where, with his sister and two brothers, he did the usual chores, including milking and barn cleaning. "In those days, milking wasn't considered part of the work; it was just a chore you did after the day's work," Gribble reminisces. "But one of the problems we always had was the cow manure, the vast quantity of it, which we had to handle by very primitive means. We just cleaned it out with shovels and forks from the barns, and if the weather was suitable, we'd put it on a manure spreader and haul it out to the fields. If the weather was too muddy to get in the fields, we'd pile it up, and then have to haul it out in the summer and spring. It was back-breaking work. Everybody hated it. If you were a little bit late to work, you knew what was going to be left for you." It was thus not entirely surprising that, as he grew up, Joe Gribble found himself drawn more toward operating tractors and milking machines than toward raising cattle.

After high school, Joe went to Oregon State College, where he studied agricultural engineering. During vacations he helped his father, Andy Gribble, in the small construction business that Andy had started after separating from Joe's mother, who kept the farm. "He was everything, the plumber, the electrician, carpenter, and cement finisher, and we got a broad education," says Joe, in his soft-spoken drawl.

Following graduation in 1957, Joe went into the Army, where he flew planes for three years. After his discharge, he went to work full time for the Gribble Construction Company. The company was essentially Joe and his father, and they put up anything from private homes to supermarkets and dairies. In the spring of 1960 they were engaged to construct a new 100-cow dairy on the Sweet Ranch at Sixes, Oregon, a small town on the Pacific Ocean near Coos Bay, two-thirds of the way down the coast from Washington. Sweet's was managed by Henry Bennett, "a very, very far-sighted individual," according to Gribble. "He was always thinking years ahead. In fact, one of the problems we had from time to time was that Hank would be worrying about the way things were going to be five years from now, and we were worrying about whether we were going to be able to pay the bills at the end of the month."

Hank Bennett had managed a number of large and successful dairies, and he carefully outlined for the Gribbles what he and Sweet wanted. Much more than a milking barn, it was to be a completely modern facility with better ways of housing, feeding, and milking the cows. An up-to-date means of washing the floor with water was to be included. This was particularly important because, as the attorney for Gribble and Bennett subsequently wrote in their "Dairy Establishment" patent application, "The removal of cow offal from a barn is as much of a task today as it was in ancient times."

For those who have never sweated on a farm, the magnitude of the manure removal task can perhaps be best appreciated from the figures an agricultural expert, Professor Milan Huber, later provided a district court when he testified as the paid expert for Gribble. Huber, author of several books on dairy housing and animal waste disposal, related that a 1600-pound cow can produce sixteen to seventeen gallons of manure and urine a day. This means, he explained, that 100 cows can leave behind as much as six or more tons every twenty-four hours. Huber said that, in 1963, many Oregon dairymen used scrapers and water hoses, some had barns with gutters and conveyors, and some still relied on the tools of Joe Gribble's boyhood travail—shovels and pitchforks.

Having followed the sagas of some other inventors in this book, you may guess what came next for Joe Gribble. He and Bennett discovered a way of removing the manure, they patented the system, and they had considerable success selling it to dairymen. Then someone else came along with a system close enough to theirs that he was accused of infringing the patent and—years later—the patent was invalidated by the Supreme Court for failing the tests of patentability.

There were eight pieces of prior art that the Supreme Court specifically cited when ruling that the Gribble-Bennett invention was undeserving of a valid patent. In order for you to see what can be used against inventors in court, let us look at those eight items.*

*The drawings on the following pages, and the majority of drawings in this book, are not photoreproductions of the originals but have been traced or redrawn to enhance clarity.

1. Systems using water to clean animal wastes from barn floors have been familiar since ancient times. As described by Witt in his *Classic Mythology,* the fifth labor of Hercules was:

> To cleanse the stables of Augeas in a single day. Augeas was a rich king of Elis, who had three thousand cattle. At night the cattle always stood in a great court surrounded with walls, close to the king's palace, and as it was quite ten years since the servants had cleaned it out, there was enough refuse in the court to build up a high mountain. . . . Hercules set to work, but even his strong arms would have failed to accomplish the task if they had not been aided by his mother-wit. He compelled a mighty torrent to work for him, but you would hardly guess how he did it. First he opened great gates on two opposite sides of the court, and then he went to the stream, and when he had blocked up its regular course with great stones, he conducted it to the court that required to be cleansed, so that the water streamed in at one end and streamed out at the other, carrying away all the dirt with it. Before evening the stream had done its work and was restored to its usual course.*

2. It has long been known that barns can be cleaned by delivering water from tanks or pools to the barn floor by means of high-pressure hoses or pipes. Hard-surfaced sloping floors in barns are also old and well known, as are various systems for introducing water in sufficient quantities and force onto the floor to wash manure into downhill drains and ditches.

3. In 1934, Frederick J. Kreutzer obtained patents for a large cow barn with a concrete floor, a milking station, and feeding stations. The milking station, which held far fewer cows than the feeding stations, was connected to them in such a way that the cows were protected from the rain while walking from feeding to milking.

---

*The original text spells the name Hercules as Heracles.

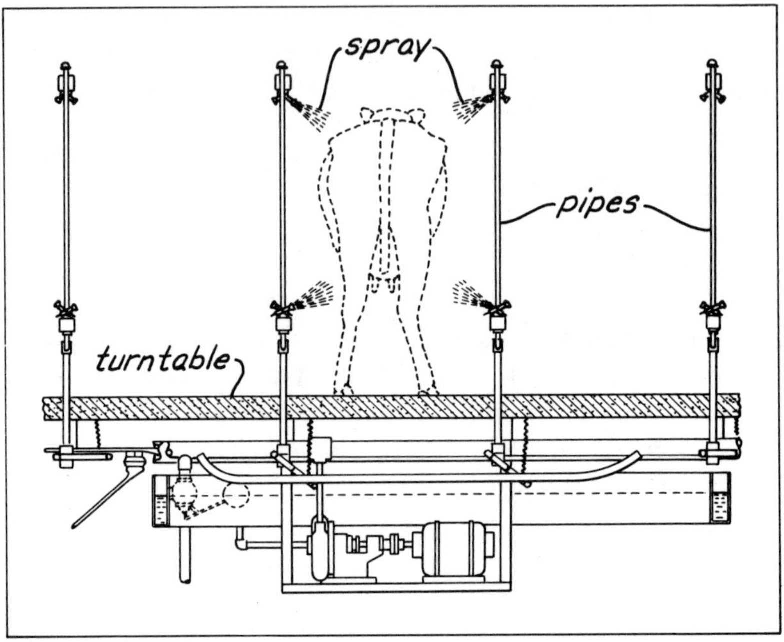

Luks' turntable.

4. That same year Kraemer Luks put his cows on a turntable. He patented a circular milking apparatus in which the cows stood around the circumference of the turntable, their heads toward the center. As the table slowly turned, they were milked and their hindquarters were spray-washed and dried. Water, under pressure, was pumped through pipes to spray-wash nozzles directed at the cows. This water drained into a gutter located below the cows' tails.

5. In 1937, Herbert McCornack patented an automated dairy plant with a separate milking area in which cows stood on an inclined bench next to shields that separated them during the milking. When the operator stepped on a foot-pedal valve, pipes directed water downward along the shields to wash them. This water then ran downhill along the inclined bench to flush off manure.

6. In 1941, John W. Bogert located two inclined gutters (troughs) under the hindquarters of rows of cows in stalls. A

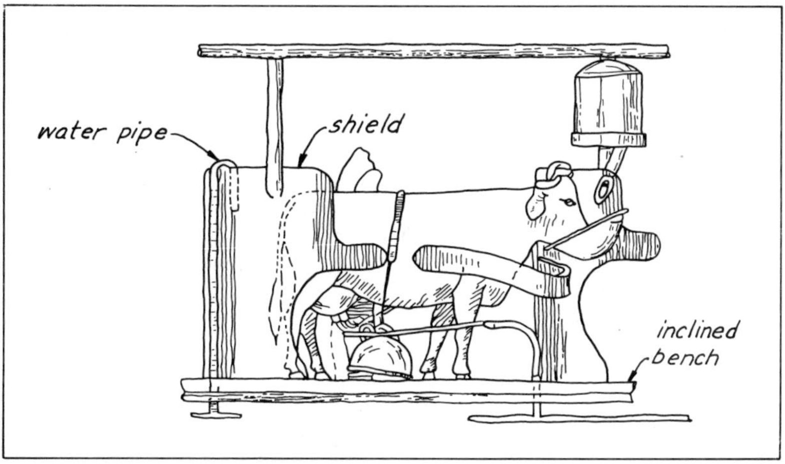

McCornack's apparatus.

walkway for the cows ran between the two gutters; buried in it were pipes through which water flowed from an elevated reservoir into spaced portions of the gutters. The rushing water cleaned the gutters of manure.

7. In 1942, Harold O.C. Ingraham patented an arrangement whereby a cow's hindquarters would hang over a depressed gutter regardless of whether the cow was standing or

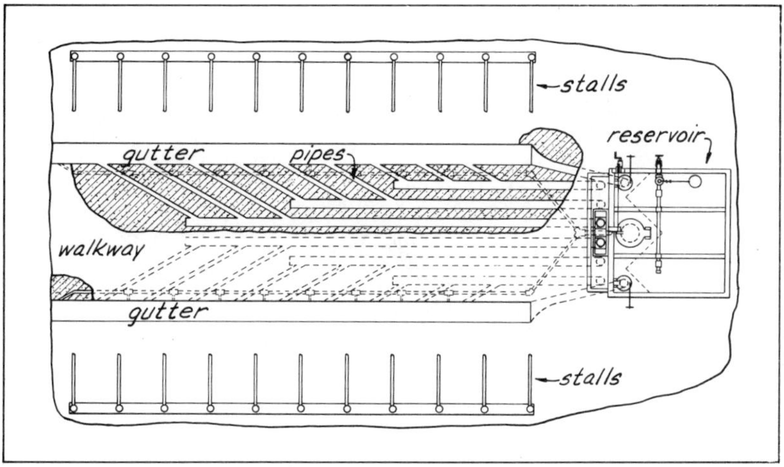

Bogert's apparatus.

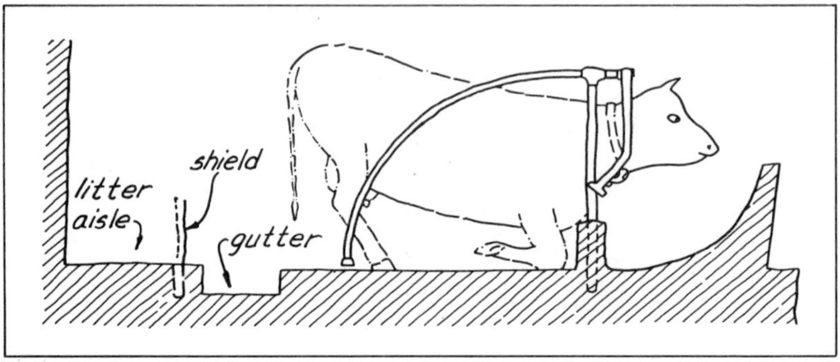

Ingraham's arrangement.

lying. A removable shield behind the gutter prevented most of the dung from spreading to an elevated litter aisle. When it was time to clean the litter aisle, the shield was removed and the litter was swept or shoveled into the gutter.

8. In 1960, a German company named Holz obtained a French patent for a barn constructed so that the cows' hindquarters were disposed over a sloping ditch or trough. The ditch was filled with stagnant water that was contained by a door at the lower end. Periodically, this door would be

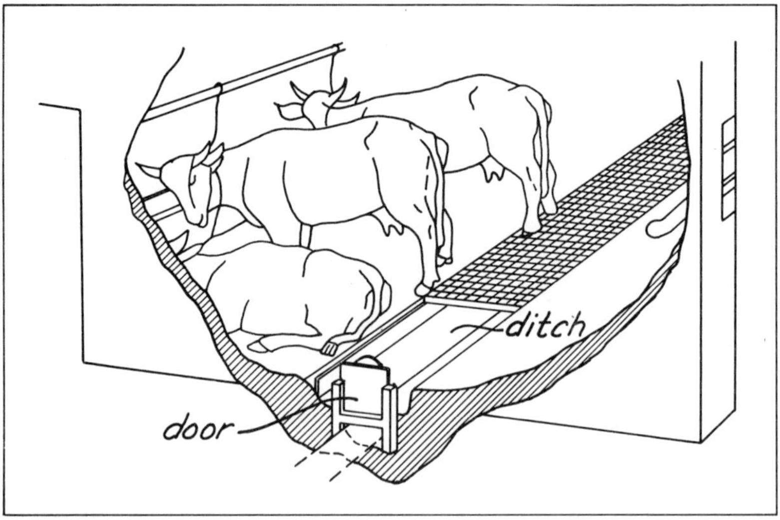

Holz's barn.

opened so that the water and dung flowed rapidly, by gravity, along the ditch to a manure tank.

Very probably, it has not occurred to you how Gribble and Bennett solved their problem. Imagine, then, how much harder — how much less obvious — it must have been for them when they could not even be certain that there was *any* practical solution. (One of the most severe handicaps faced by inventors is that they don't know whether a solution is achievable at all; they are often told it is not achievable.)

Hank Bennett and Joe Gribble spent the better part of the winter of 1960–1961 visiting dairies and agricultural experiment stations in search of information and ideas. Their first observation was that no matter how a dairy was drained and cleaned, the process was more effective when the dairy stood on a relatively steep incline — more than the usual one-foot rise every one hundred feet. Consequently, when construction began on the new Sweet Ranch dairy in the spring of 1961, their first major decision was to grade the site to a 3 percent slope (a three-foot rise for every one hundred feet). "At that time, 3 percent was just a guess," says Gribble, "but as it turned out it was a lucky guess. Since then, after building a number of these facilities, we've found out that a range between 2.5 percent and 6 percent is what works. If you go above that, it gets too steep and uncomfortable for the cows, and if you go below that, the effectiveness of the cleaning diminishes very rapidly."

Once the first buildings were erected and the cattle had been moved in, Gribble and Bennett began experimenting with different schemes for cleaning the floor. First they tried the obvious method of blasting the manure away with water from high-pressure hoses. But this procedure only cleaned the area that was hit directly. A few feet away from the point where the stream hit, the water merely dribbled over the surface, doing little good. This is the reason a hose has to be played over every square foot of floor, a tedious task that can take a person an hour and a half and several thousand gallons of water to clean 10,000 square feet of concrete. As this chore has to be done twice a day in a typical dairy (after each milking), the seemingly simple task of hosing can turn out to be an expensive operation.

After giving up on hosing, Gribble and Bennett experimented with different arrangements of sprinklers, varying the angles and locations in an attempt to find a configuration that would give the best cleaning action. But sprinklers proved to have the same drawback as hosing: each sprinkler only cleaned a doughnut-shaped area where the water actually struck the floor. And sprinklers were not cheap. So they abandoned sprinklers, and instead ran water at low pressure through various arrangements of pipes. But at low pressure the water simply skirted the piles of manure. Increasing the water pressure until it was strong enough to dislodge the manure would have required costly valves and pumps, so this idea, too, was jettisoned. Other schemes they considered, employing tractors and "scoopmobiles," were impractical because it was difficult to operate such vehicles in the midst of a herd of cows, and they did a poor job of cleaning.

Failing to obtain satisfactory results from any of the conventional barn-cleaning methods, Gribble and Bennett were forced to look for an entirely different approach. "Being right on the edge of the Pacific Ocean, why, we went down and brooded about it, looking at the ocean," Gribble recounts. "For some reason, the waves are able to keep the beach swept pretty damn clean, and, in fact, if you study where the waves pile the driftwood, it's at the extreme end of where the water runs to. So we thought that maybe we could turn loose a wave like this."

The idea of a "tidal wave" effect raised a series of questions that had to be answered: Could such a "wave" be generated every place in a barn where the cows walk? Would it clean the floor effectively? Would it be economical? And would the cows tolerate a wave swirling around their feet and not panic? (There could not be a Herculean "mighty torrent" that would sweep away the cows with the bath water.)

Gribble and Bennett generated their first wave in the most rudimentary way possible. They wedged a plank across the shallow channel running the length of the dairy barn. After sealing their crude dam by throwing manure around the edges, they let water accumulate behind the plank. Then, just as the water was about to spill over the top, Joe Gribble picked up a sledge hammer, swung heartily, and knocked out the plank. He chuckles quietly as he recalls the whole mess running down the slope so fast that "we got a hell of a good cleaning job." Very importantly, they subsequently discovered that the cows became used to their surprise foot baths and did not panic.

Having demonstrated that using a wave, or flash flood, was a sound idea, they set about building a better wave generator than a sledgehammer-actuated plank. Their first attempt took the form of a large wooden vat, nearly five feet deep and twelve feet wide, of heavy tongue-and-groove lumber. They mounted the vat on a pivot at the high end of the barn and filled it with water. Gribble's reserve melts into hearty laughter as he tells what happened when the vat was activated:

> It cleaned the barn out real good, but when it tipped over, it had so much inertia from the weight of the water that it tore itself all to pieces. We only got to use that vat one time. We just had planks all over the place. But it did a real good job of cleaning up the barn. We didn't mind so much because it was really doing the job. When I remembered my days back at home shoveling that manure out into the wagon and hauling it away, and saw that for one hundred cows we could just *squoo. . .sh*, clean it all up in about twenty seconds, why we thought we were onto something then.

Until that moment, Gribble and Bennett had been mostly "tinkering around" on weekends, their principal energies concentrated on the actual construction of the dairy itself.

Now, however, they set out in earnest to improve upon their wooden plank and vat.

They made the vat a great deal stronger, inserting shock absorbers so it would not tear itself apart after dumping the water (see middle figure in the patent drawing). They also mounted a counterweight on the back side of the vat so that the empty vessel would automatically right itself for refilling. The plank evolved into a dam or gate in the form of a steel plate, hinged at the bottom and secured to the floor, as shown by the lower drawing. Rubberized sheeting replaced the manure as a sealant. When a latch was released, the dam opened suddenly and fell flat on the floor.

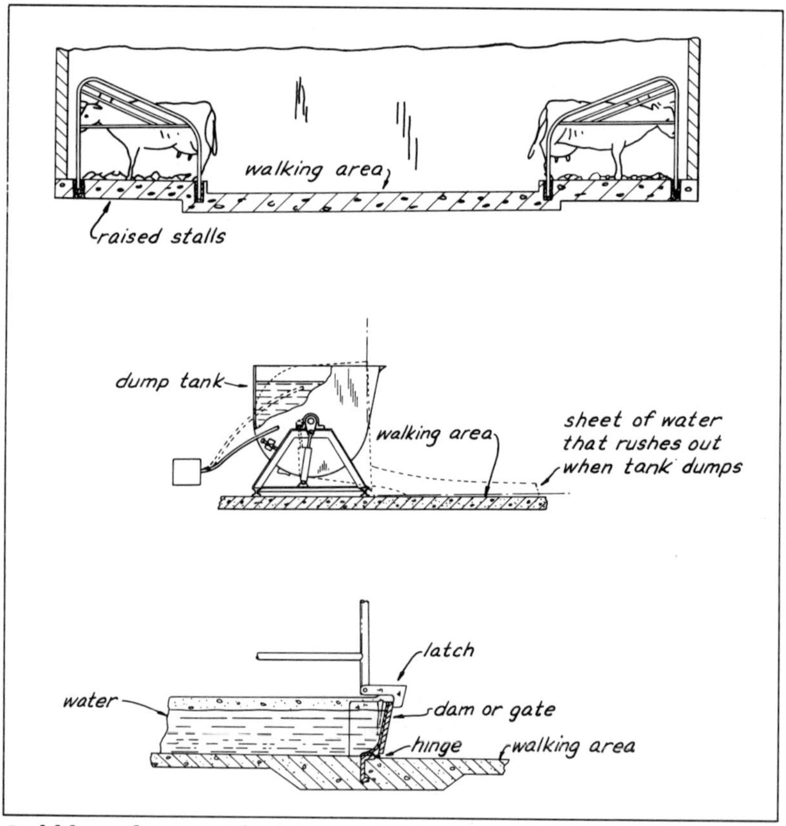

Gribble and Bennett's dairy establishment.

As described in their patent,* Gribble and Bennett allowed the cows to walk at will, on smooth inclined walking areas, almost anywhere in the barn. Also, the rows of individual raised stalls shown in the top drawing were shallow enough that the cows' behinds overhung the walkways, there being no need for troughs or gutters. (Incidentally, all the cows in a given herd are approximately the same size.) The quick-release dam would flash-flood and thus clean the walking area leading to a milking wing, and dump tanks would flash-flood other walking areas next to the elevated stalls.

The patent claims recited the combination of the inclined floor, the stall, and so forth. The key clause was:

> floor-washing means including means . . . constructed to collect water as a pool above said floor and operable after such collection . . . to disperse the water as a sheet of water over said floor.

Detailed claims called for the same combination but additionally specified (1) a dump tank adapted to tip and cascade water onto the floor to form a "sheet," and (2) the "abruptly openable" dam.

With this, Gribble and Bennett had a system that was the essence of simplicity. So simple, in fact, that in September 1971, Gribble told a United States district judge sitting in a Texas courthouse that:

> Most people didn't think it would work. It seemed too unbelievable, and this is still the reaction. If we describe this same thing verbally to a dairyman who has not seen [it] or even if we show pictures and slides of it being done, it is difficult for them to believe the kind of job that it is possible to obtain. This system, when it is properly laid

*The "Dairy Establishment" patent application was filed on November 5, 1963 and issued as Patent No. 3,223,070 on December 14, 1965.

> out, will literally clean up thousands and thousands of square feet of floor area with virtually zero labor, in just a few minutes—actually just a few seconds.

Later in the trial, Professor Huber described the "great impression" the invention had made on him when he first saw it during a visit to Sweet's Dairy early in 1963. "It was a complete new concept in manure removal, in cleaning the dairy barn, different than anything that had ever been used." Huber went on to explain that the key to the Gribble-Bennett cleaning method lay in the wave's rolling action, which resulted from the water at the top moving ahead of the water underneath. "You do not get this in a hose. You do not get it in a gutter as has been used in the past."

The highly automated Sweet's Dairy attracted a number of visitors besides Professor Huber. Perhaps the most important of these for Joe Gribble's future was Joseph B. Schaefer. He was a New Jersey electronics engineer who had turned to cattle raising when, he told the *San Francisco Examiner*, "I began to feel nervous if I hadn't told my secretary where I was going." The *Examiner* described how in 1961 Schaefer had left New Jersey and had shipped his prize-winning herd of 113 animals to Calistoga in the heart of California's Napa County wine country. Schaefer and his wife, Peggy, named their new ranch Piney Hill West and set out to build a dairy that would be a showplace. It was to have the latest dairying equipment and procedures, and it was to be the kind of place, claimed the newspaper, where Peggy could fulfill an old wish for a barn she could visit in high heels.

Early in 1963 Gribble and Bennett landed the job of building Schaefer's new dairy and installing their so-called manure-flush system. Its first commercial sale in hand, the Gribble Construction Company moved to California, where it incorporated and became AGPRO, a privately held Santa Rosa company whose principal owners were Joe Gribble, Hank Bennett, and Joe Schaefer.

The Piney Hill dairy was dedicated with considerable

fanfare in May 1964. The barn was the scene of one of the year's most unusual cocktail parties, *The Weekly Calistogan* reported (presumably, Peggy Schaefer got her wish). The paper hailed the new dairy as a "harem of 100 pampered beauties." The *Santa Rosa Press Democrat* called Piney Hill the dairy of the future, where the cow is "treated like a queen." The *Democrat* noted that the architect was Audry Emmons, who had worked on the Bay Area Rapid Transit (BART) system in San Francisco.

Hercules Goes to Court

Many inventions are difficult to copy. Would-be infringers may need complicated tooling, stacks of blueprints, sophisticated kinds of molds, or expensive machinery. Gribble and Bennett's invention, however, consisted of easily obtainable elements, such as gates, latches, and concrete. Thus, there were soon two alleged infringements of the patent. Both were later settled in favor of AGPRO without trial (a common occurrence). But then came the affair of the Jessup Dairy.

One of the first enterprises to use the Gribble-Bennett invention was Dawn Dairies in Corcoran, California. On this job AGPRO prepared the plans for construction, and supplied the equipment for the manure-flush system. A supervisor on the Dawn Dairy job was Bernard A. Sakraida, a twenty-two-year veteran in the construction business, who was given plans and drawings by AGPRO.

Subsequently, Sakraida acted as contractor or advisor for the construction of another dairy, the Jessup Dairy near Bakersfield, California. When Gribble learned of this installation, he investigated and requested that a license be taken, but none ever was. Later, on October 7, 1968, AGPRO brought an action for alleged patent infringement against Sakraida. The complaint was filed in the United States District Court for the Western District of Texas, in El Paso.* On

*The case was filed in Texas for two reasons (1) Gribble's attorney believed that Texas courts were more favorable to patents than California courts, and (2) by 1968, Sakraida was located in Texas.

November 5, 1969, Sakraida's attorney requested a "summary judgment."

In contrast to a patent trial, in which witnesses tell their stories to a judge and, occasionally, to a jury, a summary judgment is a quick piece of work by which a judge normally obtains the evidence from written affidavits rather than from witnesses testifying in person. The Federal Rules permit summary judgments when it can be demonstrated that "there is no genuine issue as to any material fact and that the moving party is entitled to a judgment as a matter of law."

Sakraida's attorney, in his brief, argued, "All that the plaintiff has done is to take a number of things which have long been known and used in the dairy business, and put them together in a form that in itself is not new, or if anything can be pointed out that may be in any way different, that it is an obvious expedient under [Section] 103."

Then Sakraida got a wonderful break. On December 8, a little over a month after the summary judgment had been requested, the Supreme Court demolished Charle Neville's road-paver patent. Sakraida's attorney lost no time in filing a supplemental argument:

> The point at issue in this case is controlled by the recent decision of the United States Supreme Court in Anderson's– Black Rock, Inc. v. Pavement Salvage Co., Inc. . . . In that case, . . . the patented item was simply an amalgamation of things already known and used, and although the product turned out to be useful, and was successful commercially, still it did not meet the legal test of being patentable.

Gribble's attorney was J. Pierre Kolisch, a Portland, Oregon, patent lawyer who had written the patent and who was to contest the case clear through to the Supreme Court. Kolisch's argument opposing the summary judgment request was of no avail. In ruling for Sakraida, the judge held that there was nothing patentable about the Gribble-Bennett "combination of . . . old elements." He then declared that the case was "controlled by the holding and opinion of the Supreme Court of the United States in Anderson's – Black Rock, Inc. v. Pavement Salvage Co., Inc., opinion dated December 8, 1969."

Gribble appealed the decision to the United States Court of Appeals for the Fifth Circuit. On February 1, 1971, three appellate judges unanimously reversed the trial court's decision. Patentability could not be determined, they stated, "until the facts are resolved, and it is apparent that the facts were not resolved at the time summary judgment was entered. This was not the 'rare patent case which can be . . . summarily truncated.' "

A remarkably quick, one-day nonjury trial took place in September 13, 1971. Sakraida's attorney, in an opening statement of about 200 words, asserted that "a combination patent is almost impossible to have because you don't get anything new," that everything about the patent was old, and that, therefore, there was nothing nonobvious about it. Then came the testimony of witnesses.

When an inventor testifies in court, he is often in a position a little like that of the victim in a rape trial. But instead of having his chastity and morals attacked by a defense lawyer, the inventor has the merits of his invention attacked. For example, here is a question Sakraida's lawyer asked Gribble:

> And you don't claim there is anything too imaginative about this dam or gate, do you, and the way it is hinged into the floor and the way this latch is up there at the top to hold it and open it?

Answer by Gribble:

> You should have been an engineer. You grasp these things in a second. It took us months to conceive and develop

> it. Yes, there was a good bit of thought and time and energy and trial and error type work that went into the development of this system as it evolved.

Later, Sakraida's lawyer drew analogies to dump trucks, drinking-water tanks, steel mills, and the use of oil and water in warfare. He asked Gribble, ". . . probably you have seen a picture of a converter in a Bessemer steel furnace, such as in a steel mill where it is filled with molten metal and it can be dumped?"

"Yes."

"That has been in use for a hundred years, hasn't it?"

"Yes, sir."

"It was in use long before that, back in Roman times when they made those things up on top of a fort and put hot oil and water in them and they would dump it out on the enemy when they approached."

In defending his patent, Gribble showed a movie in which a boy cleaning a barn with conventional hosing methods was followed by a girl who took just a few minutes to do the job with the patented apparatus. The expert testimony of Professor Huber, mentioned earlier, was followed by that of Darrell Reed, a Waco, Texas, dairyman who had installed the AGPRO system in 1969. When Reed was asked what his impression had been, he replied:

> Well, of course this was so revolutionary from the methods we had been using over the years, and like I say, I think I was kind of born on the end of a manure scoop. We had done this since childhood and the thing I think that probably impressed me the greatest was the cleanliness of the thing after the flush. This was . . . clean, clean, you know, much better than what we had been able to do with our scrapers and hoses and so forth in the past.

Sakraida's evidence included affidavits from a number of people who claimed knowledge of earlier flush systems and who judged Gribble's invention to be obvious. Only one defense witness, an El Paso dairyman, testified in person. He told of a business associate who had traveled in Costa Rica and had come across "several instances where water was diverted from a stream through a barn to clean out the offal. . . ." However, any such instances outside the United States are not prior art under the patent law.*

The judge decided in favor of Sakraida, ruling that "said Patent . . . does not constitute invention, is not patentable, and is not a valid patent, it being a combination patent, all of the elements of which are old in the dairy business, long prior to 1963, and the combination of them as described in the said patent being neither new nor meeting the test of non-obviousness, Defendant is entitled to judgment."

So, again, back up to the Court of Appeals. This time, however, that Court really went into the merits of the Gribble-Bennett invention, and on February 5, 1973, a three-judge panel ruled unanimously that the patent was valid. In reversing the trial court, the Court of Appeals took a different position on the issue of newness:

> Admittedly, all the elements are old. However, prior to this patent, the concept of a *controlled sheet of water to do the scouring and washing of the floor had not been used in barn construction.* The prior art involved the use of spot delivery of water by high pressure hoses. None of the prior art documents shows a floor washing means constructed to collect a pool of water uphill from or on the floor so that, when the water is *abruptly released,* it will flow *as a sheet* to wash almost instantaneously all barn

*Conversely, any patents or printed publications—as distinguished from uses—*are* prior art, regardless of where in the world they exist (Section 102 of the patent law).

> floor areas and discharge cow offal into drains. That is the crucial element of this patent. (italics mine)

The Court of Appeals disparaged the affidavits that Sakraida's attorney had submitted to the trial judge as being "simplistic statements, rendered without the aid of cross-examination and devoid of specific illustrations of similar inventions. . . ." It observed that Sakraida's single trial witness "admitted that the patented dam-release flooding technique was novel even to him."

In the trial court, there had been arguments about gravity, with Sakraida's attorney insisting that Gribble had simply been making use of a natural force—gravity—which, of course, no one can patent. The Court disposed of this argument, saying:

> Certainly, the gravitational principle that water runs downhill is both old and obvious; yet, plaintiff does not seek to patent gravity. Instead, plaintiff strives to acquire official recognition of its rights in a precise combination of mechanical components capable of harnessing a sudden rush of water to effectively clean a modern dairy barn. The prior art did not mother this audacious device; Gribble and Bennett's inventive faculties did.

As for the invention's "simplicity":

> Although the plaintiff's flush system does not embrace a complicated technical improvement, it *does achieve a synergistic result* through a novel combination. . . . The little expert testimony presented at trial supports the conclusion that the system was initially viewed with disbelief. The inventors . . . constructed a totally new structure (although admittedly out of old parts) and achieved a dramatic result. (italics mine)

Although the Court of Appeals thus held the patent valid in early 1973, it was not until 1975 that the trial court was finally directed to decide the issue of infringement.* This never occurred, however, because Sakraida's attorneys applied for and obtained a hearing in the United States Supreme Court. (As a result, the question of infringement—always denied by Sakraida—was never resolved.)

With the case now in the Supreme Court, the briefs became elaborate and exhaustive. In their fifty-three-page opener, Sakraida's attorneys hammered away by citing the road-paver case, A&P, and others. In his answering brief, Kolisch drew distinctions between this case and both A&P and Black Rock. He also cited a 1923 Supreme Court decision in which simply tilting a well-known papermaking machine had created a patentable invention. The patent involved in that case was issued to one William Eibel in 1907. A court of appeals had held Eibel's invention unworthy of a patent, partly because it was based on the principle of water running downhill. However, Chief Justice William Howard Taft later spoke for a unanimous Supreme Court in deciding that because no one had solved the particular problem until Eibel, despite a decade of trying, and because after Eibel announced his discovery "all adopted his remedy," there was no doubt that what Eibel had done "was not obvious and did involve discovery and invention."

The actual hearing in Sakraida v. Ag Pro, Inc., case 75-

---

*A long battle intervened about a motion for new trial filed by Sakraida's attorneys on the ground of allegedly newly discovered evidence. The appeals court directed the trial court to look into the matter to see whether there should be a new trial, which, under the Federal Rules, is only permitted relative to new evidence "which by due diligence could not have been discovered" at or near the time of the original trial. The trial court heard witnesses testify about the new evidence and then ruled in favor of Sakraida in all regards, declaring the patent invalid. Later, the Court of Appeals unanimously ruled against Sakraida on the ground that there had not been "due diligence." It did not rule on the merits of such evidence, and the evidence was not used by the Supreme Court as a basis for its opinion.

110, came on March 3, 1976, seven and a half years after Joe Gribble had first filed his complaint. Like most hearings in the Supreme Court, it was not an affair where the Justices simply sat and listened to sermonlike arguments by the lawyers. Rather, there was vigorous questioning by the Justices. There was also a good deal of laughter over lines such as "non-toilet-trained cows."

As soon as Chief Justice Warren Burger opened the proceedings, an attorney for Sakraida rose to deliver his argument, for which thirty minutes were allotted. After emphasizing that every element in the "Dairy Establishment" patent was old, the attorney fired his heaviest legal cannon: ". . . this case is controlled and governed by the decisions . . . rendered by this Court in Anderson's – Black Rock, Graham vs. John Deere, A&P. . . ."

The attorney emphasized:

> The only means that we're talking about in this particular patent that is allegedly new and outside the prior art, is the means to house or hold water at the upper end of a dairy barn in a pool, or on the floor, so that . . . it can be released suddenly, to flow downhill, and as it flows downhill the natural propensity of water removes manure that is on . . . the concrete floor.

Later, he cited Webster's dictionary as defining "flush" to mean "a sudden release or washout, sudden rapid flow of water to flood," and emphasized that "it would be most difficult to come up with a new idea that added to the sum of useful knowledge with a 'flush system.' " He then discussed Hercules' fifth labor:

> . . . I think it is relevant, to discuss this legend of Hercules. What did he do? It is recorded, it is only a legend, but it is recorded in Greek Mythology, that as Hercules, as his fifth labor, saw the conditions of the

> stables, he, at a single glance, diverted the stream, Alpheus, to clean the stables of the filthy manure on the floor.

Sakraida's attorney was then asked: "Is that the basis for your claim [of obviousness]?"

"No, sir, it is not."

"It helps, though, doesn't it?"

"It certainly does."

Kolisch, when his turn came, directed his argument primarily to the advance created by the invention:

> So what did these two inventors do? They came along and they solved a problem, and the record shows that to clean up a comparable alleyway, three or four hundred feet long, would take four or five hours. Literally, in one minute, according to the patented invention, just as long as it takes to wave my hand, you've cleaned it.
>
> Now, what you've done is you have substituted for hours of dirty, disagreeable, miserable labor, a shining clean floor that takes no time, no effort, no money, and satisfies all of the sanitary requirements. . . .

One Justice did not seem to be impressed by the difference between delivering water from a number of individual pipes or hoses and delivering it from one large wave-generating mechanism. Another Justice did not at first realize that the cows could remain in the area while the cleaning was going on. After Kolisch had distributed pictures showing the barn "before" and "after" cleaning, the following exchange took place:

> *Justice:* Well, where were the cows between before and after?

Kolisch: The cow may be in its stall, or it could be standing in the alleyway. All it does is wash its feet. As a matter of fact, the cows are quite happy to have their feet washed.

It removes the offal, and you look at me quizzically, and I don't blame you; I didn't believe it would work, either. As a matter of fact, when this thing was first presented to me, I said it's incredible, I just don't think it will work.

On April 20, 1976, the Supreme Court of the United States issued its *unanimous* decision and opinion: "We hold that the Court of Appeals erred in holding the patent valid and also in reaffirming its determination of patent validity." Thus, instead of expiring on December 14, 1982, seventeen years after its issue date, the patent terminated many years earlier. Over the eight years of litigation, two district judges had worked on the case, both ruling against the patent; eight court of appeals judges had worked on the case (mostly on procedural grounds), and all had ruled in favor of the patent; and nine Supreme Court Justices had ruled against the patent on appeal.

In its opinion, delivered by Justice William J. Brennan, Jr., the Supreme Court stated: "Systems using flowing water to clean animal wastes from barn floors have been familiar on dairy farms since ancient times." This statement was supported by the Hercules fable, which was quoted in a footnote. Like Justice Douglas' Black Rock opinion, that by Justice Brennan scrambled the statutory obviousness test with A&P concepts. But the opinion went further toward A&P than did that in Black Rock, because Justice Brennan said: ". . . the District Court properly followed our admonition in . . . A&P [that] 'Courts should scrutinize combination patent claims with a care proportioned to the difficulty and improbability of finding invention in an assembly of old elements.'. . ."

Concerning synergism, the Court of Appeals had concluded that AGPRO's combination "does achieve a synergistic result." The Supreme Court disagreed:

> We cannot agree that the combination of these old elements to produce an abrupt release of water directly on the barn floor from storage tanks or pools can properly be characterized as *synergistic*, that is, "result[ing] in an effect greater than the sum of the several effects taken separately." Anderson's – Black Rock v. Pavement Co. . . . (italics mine)

In concluding, the Court said:

> Though doubtless a matter of great convenience, producing a desired result in a cheaper and faster way, and enjoying commercial success, Dairy Establishment "did not produce a 'new or different function' . . . within the test of validity of combination patents." Anderson's – Black Rock v. Pavement Co. . . . These desirable benefits "without invention will not make patentability." Great A.&P. Tea Co. v. Supermarket Corp. . . .

The Rock-Bottom Spring of 1976

The AGPRO opinion hit lawyers and informed inventors like an electric shock. They were stunned that Black Rock's resurrection of A&P concepts had not only been continued but expanded by all nine Justices, including two who had replaced, after Black Rock, the most hyperstrict Justices (in patent cases) in Supreme Court history. (Before AGPRO, Black Rock had been regarded as an unfortunate anomaly.)

AGPRO rapidly became a cause célèbre, primarily because of the Court's use of the word "synergistic." The American Patent Law Association, which joined Kolisch in petitioning the Supreme Court for a rehearing, declared in its brief:

> While "synergism" [the word usually used instead of "synergistic"] may have a tolerably precise meaning in chemistry, nobody knows what it really means in any other application. . . . It is a figure of speech meaning something cute from an engineering point of view, or something ingenious. But as a word borrowed from chemistry, for use elsewhere, it is meaningless, an imprecise and dangerous figure of speech.

In addition to the APLA, several bar associations joined Kolisch's request for reconsideration; but the Supreme Court denied the petition. The result, as synergism became a kind of tail to the whole-must-exceed-the-sum-of-its-parts test of A&P, was revolving-door confusion in the lower courts. The confusion was magnified by the fact that, only three weeks before AGPRO, the Supreme Court had decided another test-of-patentability case without using synergism or *any* A&P concepts! In that case, Dann v. Johnston, involving a patent application claiming a record-keeping machine system for financial accounts, obviousness and the Graham approach were the only criteria applied by the Supreme Court in reversing the Court of Customs and Patent Appeals and in holding that no patent should be granted.*

Patent lawyers, when defending infringement actions, latched on to synergism as the new shibboleth and trumpeted the word to judges as if it might have magic power to destroy patents. And, frequently, it seemed that it did. Many judges stated that there was "lack of synergism" relative to any invention that they felt did not warrant a patent. Synergism joined

*I am not describing the Dann case at any length because it has not had a major impact on lower-court decisions. Lawyers close to the case speculate that there are several reasons why it was of little note. Perhaps the most important is that the Supreme Court elected not to decide what was previously thought to be the major issue — whether or not computer programs *as a class* are patentable. Lawyers then lost interest in the case, and what transpires in the lower courts depends in large part on what lawyers argue.

earlier catchwords applied to inventions that consist of "combinations of old elements."*

Although patent lawyers were upset about the synergism part of the AGPRO *opinion*, they should have been even more unsettled by the Supreme Court's *decision*, by its disregard of the strong signposts or secondary factors that were present. The answer to the problem of cleaning barns had been sought for centuries. Yet, though the elements (primarily electric pumps) necessary for achieving a practical solution had existed for many decades, the solution had not been found. By any reasonable "obviousness" standard, the Gribble-Bennett invention should have been held patentable over the eight pieces of prior art. The invention even performed the feat, uncommon for mechanical subject matter, of passing the A&P "unusual or surprising consequences" test. Finally, there was striking commercial success for such a small, new company. Before 1971, AGPRO had designed and installed some seventy-odd manure flush systems all over the United States and as far away as Japan. Customers ranged from state agricultural colleges and experiment stations to large commercial dairies such as Ralston-Purina, Dean Foods, and Foremost Milk Company. Each of the installations handled between 200 and 2,000 cows.

The AGPRO case left patent lawyers in a quandary, wondering what was required of a patent in order for it to be sustained. To them and other Americans believing in an effective patent system, the confusing opinion seemed close to the

*There is a direct, traceable "genealogy" that extends from the AGPRO case clear back to that of the coal stove. The Supreme Court's AGPRO opinion cited Black Rock, which cited A&P, which cited a 1938 aggregation case called Lincoln Engineering, which cited an 1881 aggregation case called Pickering, the latter being the first Supreme Court case to use aggregation-combination as catchwords instead of as indications of the absence or presence of cooperation. Pickering cited the pencil-eraser case, which cited the coal-stove case. The catchwords aggregation-combination mutated to whole-must-exceed-sum-of-parts, which in turn mutated to "synergism."

final straw. In the spring of 1976 the climate for invention in the United States reached rock bottom. (Only a few weeks before AGPRO, the United States Senate had passed a bill, setting forth an entire new patent code, largely written by the Antitrust Division of the Department of Justice. Much of the new code was considered to be little less than terrible by a great majority of the patent bar. The House never passed the bill, so the code did not become law, but the fact that such a bill could have gotten through the Senate created distinct trauma.) Incentive for creativity was being undermined more and more, and inventors, entrepreneurs, and patent lawyers lived in a period of increased confusion and stress.

NINE

Why the Unequal Balance?

A constitution states or ought to state not rules for the passing hour, but principles for an expanding future.

JUSTICE BENJAMIN N. CARDOZO

Why the unequal balance? What caused patent owners to so often wind up second best in court battles with infringers? Unquestionably, the most important factor was the perception of patents as monopolies and of monopolies as being against the public interest.

Patents and Monopolies

To understand the relationship between patents and monopolies, we must go back to the Elizabethan Age and the dawn of capitalism. Monarchs, such as Queen Elizabeth, granted monopolies to Court favorites for the purveying of numerous items. Hume, in his *History of England,* described the situation:

> It is astonishing to consider the number and importance of those commodities which were thus assigned over to patentees. Currants, salt, iron, powder, cards, calfskins, fells, pouldavies, ox shin-bones . . . oil of blubber, glasses, paper, starch, tin, sulphur, new drapery, dried pilchards, . . . these are but a part of the commodities which had been appropriated to monopolists. When this list was read in the House, a member cried, "Is not bread in the number?" "Bread," said every one in astonishment. "Yes, I assure you," replied he, "if affairs go on at this rate, we shall have bread reduced to a monopoly before next parliament."

Naturally, these practices generated public indignation and reaction. Finally, in 1623, the historic Statute of Monopolies was passed by Parliament as a codification of the common law. The Statute declared, "All monopolies . . . are altogether contrary to the laws of this realm, and so are and shall be utterly void, and of none effect, and in no wise to be put in use or execution." The lawmakers took care to enunciate, however, "That any declaration before mentioned shall not extend to any letters-patent . . . [pertaining to] the true and first inventor and inventors of such manufactures. . . ."

Slightly over a century and a half later, the Founding Fathers of the United States recognized the importance of stimulating creativity and invention by incorporating into the Constitution the section granting Congress the power to establish the patent system.

By any reasonable interpretation of the Statute of Monopolies, and of the United States Constitution, there is no logical or inherent "link" between patent "monopolies" and evil monopolies of the Elizabethan type. Yet, surprisingly, some judges have used the Statute of Monopolies as one justification for being hyperstrict on patents. The line of argument refers to the Statute of Monopolies in order to show that patents for inventions *are* monopolies.

What is necessary—and what people have long been trying to achieve—is a reasonable accommodation between the patent laws and laws directed against evil monopolies. As Judge Rich said:

> The greedy nature of some men is such that there are always pressures toward the creation of bad monopolies. Undue preoccupation with countering this pressure, however, has quite blurred the vision of many well-intentioned but not too well-informed people to the good . . . that monopoly power can do, properly channeled and not so proscribed as to lose its effective power. . . .
>
> To this point I have taken note that the *primary* distinction between good and bad patent monopolies is that the good patent does not monopolize something the public *already has*, so as to take something away from the public. The invention covered by the good patent must be *new*. . . .
>
> But beyond bare novelty one must go one further and troublesome step to have a sound system and keep the monopoly on the good side.

Judge Rich then described the attempts to go that "one further and troublesome step," first by the doorknob case, then by the tangle of tests, then by the obviousness section.

No less an American than Thomas Jefferson struggled mightily with the problem of reconciling the rights of the public to be protected against evil monopolies, and the operation of the patent system to promote good "monopolies." Jefferson had an instinctive aversion to monopolies but a natural empathy for inventors.

Having read every Jefferson letter that I could find on the subject of inventing and patenting, I am left with the following impression: He was (as would be expected) reasonable, temperate, and rational. Naturally, his views changed from time to time during his life. When he was a member of the three-man board that examined applications for patents, the board used some rules or tests (for example, a change of materials is not patentable!) as to what should be patentable and what should not. A comprehensive test proposed by Jefferson, and that was part of a bill he offered in 1791, was that a patent not be granted for an invention if "the same is so unimportant and *obvious* that it ought not to be the subject of an exclusive right. . . ." (italics mine)

One hundred and seventy-five years later, in its Graham preface justifying "strictness," the Supreme Court referred to Jefferson at length. Jefferson's writings, the Court said, "evidence his insistence upon a high level of patentability."

Everything, however, is relative. It would, for example, never have occurred to Jefferson to treat the inventors of mechanical inventions as second-class citizens. In a letter to a Mr. Pictet, written October 5, 1803, Jefferson said, "In the arts, and especially in the mechanical arts, many ingenious improvements are made in consequence of the patent-right giving exclusive use of them for fourteen years." Of course, in those days mechanical inventions were far more numerous than chemical, while electrical inventions hardly existed. But there is no justification for discrimination after the chemical and electrical arts proliferated.

Tragically, the technical disclosures of almost all of the patents personally examined by Thomas Jefferson and his two co-cabinet officers were destroyed by fire in December 1836. There remain, however, lists of early patents. One list shows that most of the patents issued by Jefferson were for machines, methods, and the like. However, in August 1790, Joseph S. Sampson was granted a patent for "Candles." In August 1791, Ludwick C. Kuhn received a patent for a "Bedstead," while Henry Voight was granted one for "Propelling Boats By Animal Power." Samuel Green obtained a patent in March 1792 for "Hose, Made of Canvass."

Any or all these inventions may have been ingenious; we will never know. What we do know is that during Jefferson's tenure as an "examiner," there were more than twice as many patent applications filed as there were patents granted. This degree of strictness was actually liberal compared to the hyperstrictness exhibited by the mid-twentieth-century Supreme Court. On the other hand, it was distinctly more strict than the Patent Office is now.

When Jefferson and the other two cabinet officers who were members of the board became swamped with patent applications, they had to abandon the task of examination because of higher duties. For the next few decades there was no examination at all—a highly undesirable situation. Commenting on the new turn of events in a letter to Isaac McPherson of August 13, 1813, Jefferson said:

> The whole was turned over to the judiciary, to be matured into a system, under which every one might know when his actions were safe and lawful. Instead of refusing a patent in the first instance, as the board was authorized to do, the patent now issues of course, subject to be declared void on such principles as should be established by the courts of law. This business, however, is but little analogous to their course of reading, since we might in vain turn over all the lubberly volumes of the law to find a single ray which would lighten the path of the mechanic or the mathematician. It is more within the information of a board of academical professors, and a previous refusal of patent would better guard our citizens against harassment by law-suits. But England had given it to her judges, and the usual predominancy of her examples carried it to ours.

From this it may be inferred that Jefferson wanted good examinations, to protect citizens from unjustified lawsuits, as distinguished from invalidations by judges not versed in technology and invention. (What modern-day America needs is examinations as good as can be had *at reasonable cost,* coupled with court actions presided over by judges who thoroughly understand the patent system and the inventive process.)

Besides Jefferson, two other great Americans were concerned with patents and inventions in the early decades of the existence of our country. The first was Benjamin Franklin. In his autobiography, Franklin wrote:

> Gov'r. Thomas was so pleas'd with the construction of this stove . . . that he offered to give me a patent for the sole vending of them for a term of years; but I declin'd it from a principle which was ever weighed with me on such occasions, viz. *That, as we enjoy great advantages from the inventions of others, we should be glad of an opportunity to serve others by any invention of ours; and this we should do freely and generously.*

However, Franklin's generosity, if generally applied, would undermine the Constitutional purpose of promoting the progress of useful arts. A noted electrical engineer and inventor, Elihu Thomson, who was a contemporary of Edison, described the situation with accuracy:

> I have known some well-meaning scientific men to look askance at the patenting of inventions, as if it were a rather selfish and ungracious act, essentially unworthy . . . [but] a patent for an invention enhances its commercial possibilities as well as offers the opportunity of financial reward to its originator. Ordinarily no manufacturer or capitalist would be willing to-day to risk his money, and expend time and energy in developing on a commercial scale a new product or process without being assured that his investment in developing the invention would be protected in some measure against possible pirates.

The second great American was Abraham Lincoln. Though not generally regarded as being ungenerous, he obtained, on May 22, 1849, Patent No. 6,469 for an invention in "Buoying Vessels Over Shoals." Lincoln was very interested in discoveries and inventions, and even gave a speech on the subject at Springfield, Illinois, in 1860.

Misuses of Patents, and Antitrust Violations

That the monopoly stigma and the perception of patents as monopolies had an enormous influence on judicial decisions is clear. Those judges with the most jaundiced views of monopolies have generally been the ones hyperstrict in their interpretation of the patent laws. In the mid-century Supreme Court, the government had a good record of winning antitrust cases. At the same time, an inventor or patent owner hardly ever won a patent case in that Court.

It is the Achilles' heel of the patent system that although there is no logical link between patents and antitrust violations (whether or not the violations created monopolies), many greedy men forged such a link in the first half of the twentieth century.

An early, classic Supreme Court case relating to misuse of patents, which is closely akin to antitrust, involved the nascent film industry. In 1908, the four principal companies that manufactured motion picture equipment and supplies, and produced films, ended a decade and a half of cutthroat competition and patent pirating by pooling their patents in the Motion Picture Patents Company. By means of outrageously drawn patent licenses, the MPPC tried to control the industry both horizontally and vertically, from the manufacture of equipment to the exhibition of the films. Producers could not make pictures more than two reels in length and had to have their outputs censored by a blue-blooded committee. Goon squads were employed to break up the productions of filmmakers who refused to knuckle under. In 1916 the Supreme Court ruled that the licensing activities of the Motion Picture Patents Company were illegal.

There were, during and after the time when the Motion Picture Patents Company flourished, various excessive and unwholesome uses of patents in some other burgeoning industries. One of the worst involved the glass container manufacturing industry. Supreme Court Justice Owen Roberts described the situation while writing an opinion, in 1945, declaring that the antitrust laws had been broken:

> Hartford, with the technical and financial aid of others in the conspiracy, had acquired, by issue to it or assignment from the owners, more than 600 patents. These, with over 100 Corning controlled patents, over 60 Owens patents, over 70 Hazel patents, and some 12 Lynch patents, had been, by cross-licensing agreements, merged into a pool which effectually controlled the industry. . . .
>
> The District Court found that invention of glassmaking machinery had been discouraged, that competition in the manufacture and sale or licensing of such machinery had been suppressed, and that the system of restricted licensing had been employed to suppress competition . . . and to maintain prices of the manufactured product.

Unlike such abuses based on issued patents, another abuse involved *obtaining* patents unfairly. Some patent lawyers took the position that a patent application was an adversary process requiring "gamesmanship." So, even though they or their clients knew of prior art that would, if the patent examiner became aware of it, result in the rejection of a patent application, they did not inform the examiner of the existence of that art. Thus, if the examiner failed to discover the art, when making his own search, the patent would be granted.

When some of these cases eventually wound up in court, they tended to bring the entire system into disrepute. The landmark case is Walker Process v. Food Machinery and Chemical Corporation, decided by the Supreme Court in 1965. Food Machinery had sued Walker for patent infringement, in response to which Walker counterclaimed that Food

Machinery had "illegally monopolized interstate and foreign commerce by fraudulently and in bad faith obtaining and maintaining . . . its patent . . . well knowing it had no basis for . . . a patent." (The reported cases do not show that Walker's claim was ever proven; what is quoted is a mere allegation.) The Court established new precedent by declaring that, if proven, Walker's claim could constitute an antitrust violation, provided that other requirements of such a violation were present.*

Following the Walker decision, patent lawyers and inventors became extremely careful to inform the Patent Office of everything they knew, in order to eliminate the danger of being charged with antitrust violations. Nevertheless, defendants in patent cases began *routinely* charging "fraud." Said one lawyer in 1974:

> Ever since *Walker Process* in 1965, it has become virtually standard practice for defendants in patent infringement litigation to plead fraud on the Patent Office on the part of the patentee or his representative and it has become a part of the regular "boiler-plate" in defendants' answers in such cases. Because of these indiscriminate and often frivolous accusations of fraud upon the Patent Office, the Courts have understandably taken a rather jaundiced view of the defense and uniformly rejected it, with the exception of the few cases in which there was solid evidence of conscious, deliberate, and affirmative fraud on the part of, or on behalf of, the patentee.

In 1978 a district judge expressed surprise at the frequency of the claim: "We are struck by the uniformity with which the alleged infringer defends his honor. He engulfs the court in a shower of prior art which the examiner, with all his

*In other words, fraud by itself would not be an antitrust violation unless there were accompanying factors necesary to establish monopolization under Section 2 of the Sherman Act.

search facilities and expertise, has somehow missed and, of course, points an accusatory finger at his impeccant adversary and plaintively cries 'Fraud on the PTO' [Patent and Trademark Office]. It was not until my fifth patent case that I learned that, when an accused infringer's attorney charges 'Fraud on the PTO,' he does not really mean the other fellow is a scoundrel. Indeed, I came to the wisdom that 'Fraud on the PTO' is not even a pejorative. It is a greeting, a salutation, in the patois of the patent bar."

Even more of a cliche are the persistent reports that patents have been deliberately suppressed by major companies in order to eliminate competition and maintain high prices. These reports are—unlike the actual abuses just described—a bum rap. In an old motion picture titled *Night Key,* Boris Karloff was cast as a penniless inventor victimized by an unscrupulous businessman. Immediately after Karloff and the businessman affix their signatures to a patent-license contract, the dialogue proceeds something like this:

> *Karloff:* Now you can put my system [a citywide burglar alarm] in operation at once.
>
> *Businessman:* Maybe some day.
>
> *Karloff:* But the contract said you have to proceed with commercial exploitation.
>
> *Businessman:* It doesn't say *when.*
>
> *Karloff:* But you stole my first invention years ago—became rich while I am poor. This second invention is my only chance. I just want it for my daughter.
>
> *Businessman:* It would cost us a lot of money to put in this new invention. Your old one works fine. I just bought the second invention so nobody else could.

Now, that's real suppression! But it is also fiction. Though there may be rare cases of deliberate suppression in the business world, they are very much the exceptions that prove the rule. Sociologist S. C. Gilfillan wrote: "The gross and harmful suppression of inventions is essentially a myth. . . ." Gilfillan cited authority that, of the 40 percent of patents that are assigned to corporations but never commercialized, 41 percent are left fallow because they are not competitive, 31 percent because there is no market demand, and 15 percent because they are in the process of becoming obsolete.

The links forged between patents and antitrust were long ago broken by the Antitrust Division (especially under a famous Assistant Attorney General, Thurman Arnold, in the 1930s and 1940s), by the mid-century Supreme Court, and by the 1965 Walker decision about "fraud on the Patent Office." But, although the links were broken *in fact*, they lingered on *in the minds* of many judges, to the detriment of patent owners.

Misunderstandings About the Patent System, and Lack of Understanding of It

As was written long ago by Dr. Vannevar Bush, who was the director of the Office of Scientific Resource and Development during World War II: "Judges, by and large and with notable exceptions, just do not understand what the patent system is all about."

In 1971, former Supreme Court Justice Abe Fortas described the situation more vividly:

> Many federal appellate judges—perhaps most of them—approach patents with the kind of suspicion and hostility that a city-bred boy feels when he must traverse a jungle full of snakes.
>
> The patent system is strange and weird territory to most judges. They have never seen anything that resembles it. All patents look more or less strange and threatening to them; and since they are heavily armed with the power of the U.S. Government, they frequently get the idea that it's their duty to kill everything that moves in this dangerous land.

Because of their hatred of monopoly, it goes contrary to the fiber of many federal judges to hold a patent valid and then to issue an injunction preventing the infringer from staying on the market and competing with the patent owner. Seldom do such judges ask themselves whether *either* of the products (the one that was patented or the one that has infringed) would be on the market if there were no patent system. If that system had not stimulated the patent owner to invent and market, there would often have been nothing for the infringer to copy.*

The prevalence of copying in America was described in a 1980 article titled "Innovate or Replicate?" which said: "Especially in the hotly competitive U.S. marketplace, replicating a rival's new product is as natural as a knee-jerk response to a physician's tap. Gillette brings out its Trac II dual-edged razor blade, and Schick is quick to respond with its Super Ultrex." There are, of course, many situations in which such instant copying is proper, as when the copied product is not patentable or patented. However, when the desire to permit copying—and thus continued market competition in the very products being fought over—influences a judge to declare a patent invalid, the judge obviously has a misunderstanding of the system.

A judge with accurate, full knowledge of the patent system would think: Let all of those things that aren't patented be copied instantly, but let those products that are correctly patented be copied either after the patent expires or by being "invented around" or "designed around." The concept of inventing around or designing around involves creating a product that will do much the same thing as the patented item but that does not infringe the patent. It was recently described: "The existence of claims to a limited right to exclude

*Judge Rich once wrote: "Paradoxically, in the working of the patent system, monopoly often promotes competition. Numerous instances are all around us, wherever two products serving the same general purpose have achieved, with the aid of patent protection, commercial production. And only after such production is there any possibility of market competition."

encourages others to try to come up with a better way—'invent around'—more often than not using the patented disclosure as a starting point. . . . As far as a potential infringer is concerned, the greater the perceived value of the claimed patent rights, the greater the requested royalty and the greater the incentive to invent around, thereby adding yet further to the state of the art on which still others will build."

How much better for a company, when attempting to capitalize on a market opportunity opened by a rival, to do some research of its own instead of merely coming forth with a "me too" product and an accompanying media blitz. When the second-comer attempts to design around a patent, as distinguished from replicating, the resulting product is sometimes "second best" but is often—and in my experience, usually—better.* Is it more beneficial for the public to have a choice of different products or only of different advertising and cosmetics?

Sometimes the invention is so revolutionary (what patent lawyers call "pioneer") that the patent can't be designed around. It is then that the patent system operates in its best mode. The public is given a new technology and often is given a new or greatly enlarged company that may survive and compete decade after decade in the face of competition by those long in the field. The basic patent expires, and competition moves in, but the company is still there—just as Ransburg Corporation remains healthy although its basic patents have lapsed.

Some persons who do not understand the patent system actually think that none is necessary, even though every major industrial country in the world (including Russia) has

*In a survey that included "designing around," Richard F. Carr found that 67 percent of those responding had made attempts to design around the patents of others. Of these, 61 percent obtained results they believed to be superior to the patented item; an additional 26 percent obtained items they thought equal in merit to the patented invention, while only 13 percent came up with products considered inferior. Furthermore, 75 percent indicated that their efforts to design around had opened up new fields of research or development.

one. Such people seem to believe that—even without the sytem—inventions would be made because they fill a need, or because inventors have some inner drive or compulsion to invent, or because there is an inevitable march of technology unrelated to any particular individual, or because inventors invent for the fun of it. Of course, to some extent all of these are true, or no inventions would have been made before patent systems came into existence. The key questions, however, are: How many inventions? How soon would they be made? How important are they? Would they be commercialized so that the public would receive their benefit?

One of the clearest examples of how the patent system fostered the making and commercializing of a tremendous invention is provided by xerography. Chester Carlson was a physicist-turned-patent-lawyer who worked in the patent department of P. R. Mallory and Company. Being a patent lawyer, he was able to do a good job of writing his own patent applications after years of study and experimentation (some of it in his kitchen, a la Bert Adams) produced the discovery that made him go down in history as one of America's greatest inventors.

After Carlson produced his first, barely readable xerographic image in 1938, he made at least eighteen unsuccessful attempts to interest corporations in his invention. (It "was met with an enthusiastic lack of interest," he said.) The turn-down list included such illustrious names as IBM, RCA, Remington-Rand, and General Electric. It was not until 1960 that the practical, automatic, push-button xerography we now know was completed. What sustained Carlson during years and decades of discouragement? The answer is provided by this statement he made when receiving, in 1965, the first "Inventor of the Year" award from George Washington University's Patent, Trademark, and Copyright Institute:

> The time scale of invention is a long one. Results do not come quickly. Inventive developments have to be

> measured in decades rather than years. It takes patience to stay with an idea through such a long period. In my case I am sure I would not have done so if it were not for the hope for eventual reward through the incentives offered by the Patent System.

During the decades required to bring xerography from first crude image to push-button reality, major contributions were made by companies and investors who needed patent incentives to spur them on in the vast and risky effort. The first company to make a deal with Carlson was Battelle Memorial Institute, a nonprofit organization self-described as "in the business of developing and improving technical inventions and in selling of the rights thereon *when patented.*" (italics mine) Battelle made major improvements on Carlson's invention. It then made unsuccessful approaches to thirty-six companies.

Finally, in 1946, the small Haloid Company of Rochester, New York, contacted Battelle and offered to negotiate an agreement. Thus started what a federal district judge later termed "one of America's foremost episodes of corporate success." Xerox Corporation—as Haloid was later named—was led by Joseph R. Wilson, Jr., a brilliant, farsighted, and courageous businessman. The great problems involved in further research and development, preparation for mass production, distribution, and (of course) money raising would have taxed the resources of a company that was huge to begin with. As recently as 1962, Xerox Corporation had a negative working capital.

Carlson, Battelle, and Xerox created the second greatest revolution in the history of automation of offices. The first was created by the men who invented and commercialized typewriters of the general kind we know today (as distinguished from fascinating but usually impractical machines that were developed over the course of a few centuries). The invention and development of that typewriter resulted, as did xerography and numerous other major inventions, from the relationship between a primary inventor, an original backer, and an established company capable of achieving mass production. By coincidence, the original backer of the typewriter was a lawyer (as was Chester Carlson) and wrote the patent applications (as Carlson did).

James Densmore, the typewriter backer, actually bought an interest in the typewriter *sight unseen* upon receiving a typewritten letter from the primary inventor—Christopher Latham Sholes. He then encouraged, financed, and browbeat Sholes and Sholes' co-inventors over a period of years, during which some forty different typewriter models were produced and an unsuccessful attempt at market introduction was made. Finally, Remington was contacted because it was already mass-producing sewing machines and steam plows, as well as the guns for which it was (and is) famous. (Remington did not originally buy rights to the typewriter inventions and patents; instead it agreed only to production engineer and manufacture the typewriter when paid by the Densmore group, which would do the selling. Only after the Densmore group failed in the sales effort, due largely to public apathy, did Remington take a license under the typewriter patents on a royalty-paying basis.)

Densmore's desire for wealth and his passion for maintaining as much ownership as possible in the invention are shown by a letter typewritten by him, a portion of which is reproduced here. The handwritten material, from a letter written by Sholes, shows the discouragement often felt by that great inventor. (Densmore's efforts eventually produced wealth and the honor of having his name appear—along with those of Sholes and Sholes' co-inventors—on a wall of the Smithsonian, a rare occurrence with respect to backers.)

Remington is no longer in the typewriter business, having joined several old-line American typewriter companies who have either quit or have permitted their names to be placed on foreign-made machines. Xerox, on the other hand, is

I BELIEVE IN THE INVENTION, FROM THE TOP—MOST CORNER OF MY HAT TO THE BOTTOM—MOST HEAD OF THE NAILS OF MY BOOT—HEELS, AND IT IS SUCH AN ABIDING CONVICTION WITH ME THAT NOTHING CAN DISSIPAT-E IT EXCEPT THE DIRE EXPERIENCE OF EXHAUSTIVE TRIAL AND ABSOLUTE AND UTTER FAILURE.

SO THOROUGH IS THIS CONVICTION WITH ME THAT IF I HAD TEN THOUSAND DOLLARS I WOULD NOT TRY NOR DESIRE TO SELL ANY PART OF IT TO ANY BODY, BUT WOULD INVEST IT ALL IN THE ENTERPRIZE AND GO ON WITH IT ALONE.

BUT STILL, ALL HUMAN JUDGMENT IS NECESSARILY FRAIL, AND I MAY BE MISTAKEN AND THE INVENTION MAY PROVE TO BE WORTH NOT ONE CENT.

Part of a letter typed by Densmore. Used by permission of Priscilla Densmore.

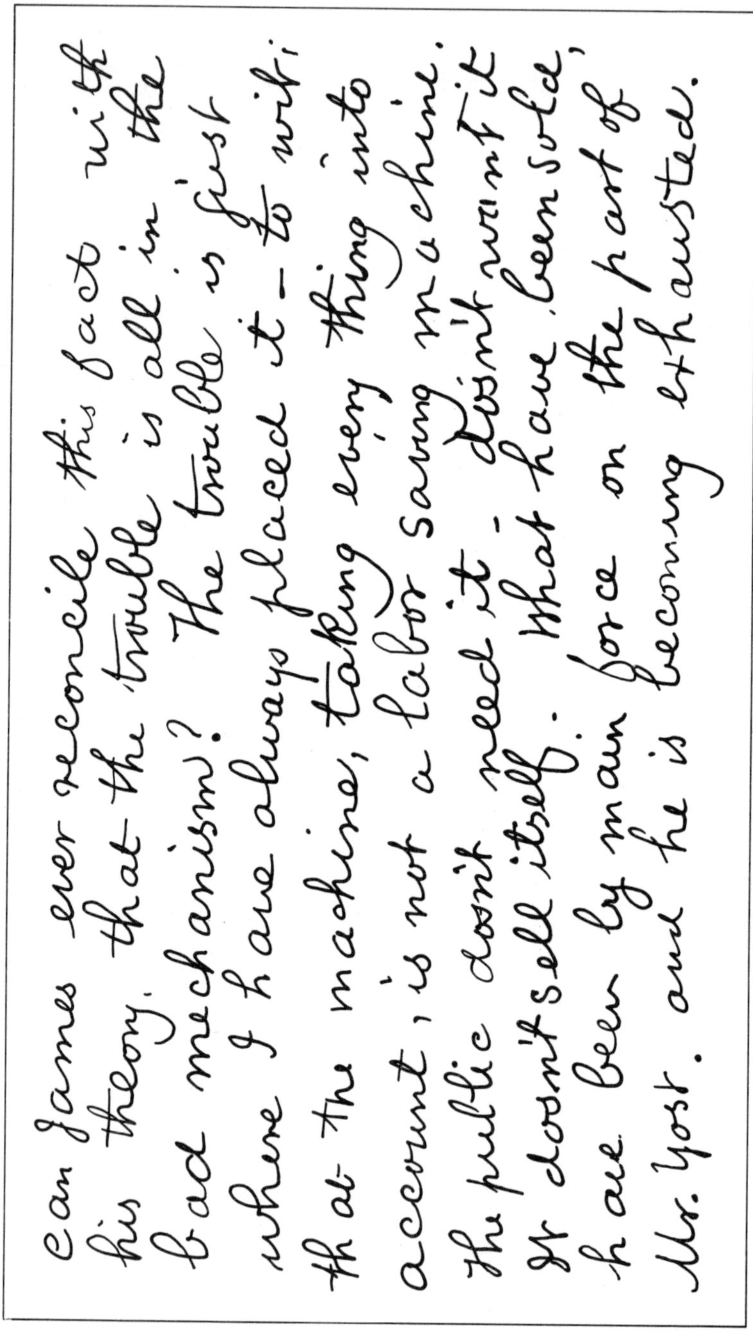
can James ever reconcile this fact with
his theory, that the trouble is all in the
bad mechanism? The trouble is first
where I have always placed it – to wit;
that the machine, taking every thing into
account, is not a labor saving machine.
The public don't need it – don't want it
It don't sell itself. What have been sold
have been by main force on the part of
Mr. Yost. and he is becoming exhausted.

Part of a letter by Sholes. Used by permission of Priscilla Densmore.

heavily into typewriters, giving competition to other giants, primarily IBM, in the field of sophisticated word processors employed in offices. What is the relationship between the patent system and huge companies such as IBM and Xerox?

A major and distinctly harmful misconception, held by some, is that big business should not have the full benefit of the patent system. A classic debate relating to patents vis-à-vis big business occurred in the late 1930s between Conway P. Coe, a prominent Commissioner of Patents of that era, and Thurman Arnold, the famous assistant attorney general in charge of the Antitrust Division:

> *Mr. Arnold:* Nevertheless, performance of these patents indicates at times that some Nation-wide corporation by getting hold of one patent or by borrowing others, or by their research, do maintain themselves in a monopoly position by virtue of the patent privilege. May I illustrate that by another hypothetical example? I am now wearing nylon socks and nylon tie, and I think nylon is entitled to this advertising after all the trouble the Antitrust Division has caused them. So I want to say that apparently they are indestructible. The socks . . . haven't worn out yet, and show no tendency of wearing out; and the tie springs back into place and you can wear one tie about 7 or 8 months. There is a possibility, therefore, that nylon may become sufficiently important as to displace the entire silk industry, the entire hosiery industry, and if that happens then one of our largest corporations will be in complete monopoly of what has got to be a necessity. . . .
>
> *Mr. Coe:* General Arnold, may I give another hypothetical example? Suppose tomorrow an inventor came along with a material much superior to nylon, then what would be the situation?
>
> *Mr. Arnold:* Then the problem would not arise, but if he doesn't, then nylon is in that situation.

Mr. Coe: The only other observation I would make on that is this, who was responsible for the invention and development of nylon?

Mr. Arnold: The Du Pont Co.

Mr. Coe: And was the public deprived of anything that they previously had by virtue of that development?

The debate seems almost quaint today. The patent on nylon has long since expired, and other fibers have been developed as competitors. Undoubtedly the silk industry was adversely affected, as were the other natural fibers: wool, cotton, and flax. But it is difficult to conceive of what the fabric industry would now be like without the Du Pont pioneering.

It is vitally important for America that big corporations be allowed to use patents fully and effectively. The importance is now greater than ever, when highly aggressive, government-helped, innovative foreign corporations are taking over many American markets. In contrast to the situation in some countries, decisions by large companies in America are not made by committees. Nor are they made by persons who own or control the companies. Thus, when a decision is made to put a new product on the market, and the product goes bad because of the high risk involved, the executives who took the risk are often in positions where their careers may be severely retarded. The problem was described in 1965 by Daniel De Simone, director of the Office of Invention and Innovation, National Bureau of Standards:

> Moreover, since the risk is great and the stakes large, a single error of judgment can seriously jeopardize the career of a corporate executive who, in his years of greatest responsibility, is likely to be between 45 and 65 years of age. That is not a time for great mobility. Where does he go after he has taken the great gamble and lost? Is

> it any wonder, then, that he has a strong tendency to avoid embracing new, untried, disruptive inventions. The consequent burden on the outside inventor (and, to a large extent, the corporate inventor, too) is to prove beyond a reasonable doubt that, in the end, there will be success. That takes some doing. We saw how IBM turned down the option on xerography, because it was too risky. If it was too risky for IBM, imagine how risky it was for the tiny Haloid Co.

Without a strong patent system to prevent copyists from riding on the backs of American corporations after their innovative gambles achieve success, executives of such corporations will be more likely—instead of taking the gambles—to think: Nothing ventured, nothing lost, including my job.* Nevertheless, the modern-day American patent system has been criticized by some on the theory that it is for the benefit of large companies, while individual inventors have been overwhelmed. It is not emphasized by these critics that there are two kinds of large companies: *creative* and *uncreative*. In 1959, Polaroid's Dr. Edwin H. Land said in reference to the latter:

> To a large and fundamentally uncreative company, patents are a threat and a nuisance, and you will find only the weakest support for, amounting almost to attack upon, the patent system, in massive corporations of this character. . . . Today . . . you will find such companies . . . disclaiming any selfish interest in building strength through invention. Now, I am certainly not suggesting that there is anything evil about institutions which are effective because of superb manufacturing techniques

*That many American corporations do less major innovating than formerly is generally recognized; the reasons often given for this state of affairs include—correctly—high interest and inflation rates.

> combined with well-planned distribution. I am, however, urging that for such companies innovation elsewhere represents a dangerous threat rather than a wholesome challenge, and that it is in the short-term interests of such companies to have a very weak national patent system so that innovation of great significance will not thrive, and also so that they may acquire at low cost the results of any innovation elsewhere, if it does become significant.

Clearly, the patent system is not for the benefit of uncreative corporations.

To whatever extent individuals have been overwhelmed, it has been due to a patent system excessively weak, slow, and expensive for effective use by them. But even with these severe defects, many individuals have started small companies that endured. Said Dr. Bush, in 1957:

> I have had personal contact with dozens of these small companies which are competing successfully, in very many cases, with large, established industrial units. They can do so because of their flexibility and specialization. But very many of these new ventures depend for their very life upon the patent system to protect them in their infancy against cutthroat competition or the weight of enormous companies in their field. Ever since this country was founded, new industrial units have thus come into being, to succumb or to be successful. They have been absorbed by larger companies very frequently, but some have grown to be large themselves. Every company in this country was once small. Many got their start because they offered something new which the public wanted and would pay for, and because the patent system held an umbrella over their heads until they could stand the full rigor of the industrial climate.

The solution is not to tear into patents or large corporations but to do more to make the system stronger, less expensive to use, and less slow-moving. This would benefit all creative individuals and their backers, and all creative corporations and their investors, so that the progress of useful arts would be better stimulated.

Misunderstandings and Lack of Understanding About Inventions

Inventors start out facing twin barriers: "The idea can't be any good or it would have been done already" and "It can't be done." They end hearing the statements: "Somebody did it before" and "It was easy."

For example, the fact that Edison was the inventor of the phonograph (which, incidentally, comes as close as anything could to being a marvelous "flash of genius") is not universally recognized. In a book published in 1976, the question was raised whether or not a French prodigy named Charles Cros was the inventor, even though what Cros conceived was never built and thus never made a single sound. As another example, Sholes' invention of the typewriter has been debunked in a highly scholarly work. An excellent book, published in 1973, lists no less than twenty-five early typewriter inventions. The name Sholes appears nowhere on that list, not even as being the inventor of the first commercially successful typewriter, such honor being reserved for a Danish inventor named Hansen who invented a "writing ball" commercialized in Europe and exported to numerous countries. Lest it be thought that the "writing ball" was analogous to IBM's Selectric typewriter, it was not. It looked, instead, like the upper portion of a relatively large sphere, with plunger-keys sticking out along numerous radiuses.

Because of the inherent characteristics of inventions, people looking at them after they were made do not have 20/20 hindsight. They have, instead, 20/10 hindsight. The resulting problems for inventors are compounded by the absurdity—but a true absurdity—I will now describe. *Logically,* the *type* and the *size* of a change (by which an invention in question differs from prior art) should have some relationship to the issue of whether or not the particular invention in question would have been obvious. But, ridiculous as it may sound, this logic does not apply at all.

Relative to the *type* of change, reference is made to this statement in an opinion by a court of appeals:

> The drain slot . . . forms the basis of this invention. . . . In other words, his patent is built upon a foundation even less substantial than sand. It is constructed upon a slot—in other words, upon a hole. Such a patent cannot stand. At most, it "is merely a mincing step forward."

One need look no further than his or her own body to find that each part, whether it be solid or an aperture, has an important place in the scheme of things.*

Relative to the *size* of the change (vis-à-vis prior art), Judge Hand said, "Nothing is more illusory, as nothing is more common, than to assume that [patentability] can be measured objectively by the magnitude of the physical readjustments required." The fanciful cartoon is representative of the way many actual inventions occur. Rounding the corners of the caveman's "sled runners," though slight in fact, would be monumental in result.

There have been numerous great inventions in which the *physical* differences between the invention and the prior art were slight indeed. As was written in 1948:

> A matter of utmost concern to all persons interested in a strong patent system is the general impression of the Federal Courts that all of the great and startling inventions belong in historical limbo and that current advances in the art, submitted as inventions in patent

*". . . the parts of the body that seem to be weaker are indispensable. . . ." 1 Corinthians 12:22.

"D'yer ever feel you're on the verge of an incredible breakthrough?"

litigation, are minor, uneventful, evolutionary increments in the steady march of technical progress.

The fallacious impression is based on a preconceived notion that the truly great inventions of history have "burst full-blown from Athene's brow" during some incandescent flash of the inventor's genius and that all true inventions are born by a similar parthenogenetic process.

As a matter of fact, the record establishes that the majority of the great and classic inventions which have been tested in the Courts, were hemmed about by the near inventions of many others and that, in each case, they represented that small advance which converted failure into commercial success. [Specific detailed examples were then given.]

It is logical, but wrong, to think of large changes in esoteric parts as indicative of unobviousness and patentability. Similarly, although it is logical to think of *technological* and *complex* subjects as being more unobvious (patentable) than prosaic and simple subjects, there is no rational reason why this should be so. A choice example of such objectionable thinking was given in the insecticide spray-cap case by lawyers arguing on behalf of the alleged infringer:

> We submit that, in a century in which man has unlocked the secrets of the atom and in which travel to the moon lies at his feet, it is a perversion of language and law to call the reshuffling of the elements of a bottle cap an invention.*

That says it all. On the one hand we have the scientists standing in front of the bank of computers, and on the other we have a person standing in some shop screwing around with a bottle cap. How could what the latter produces be patentable? The answer is simple: the patent law says—and should say—something is patentable if unobvious "to a person having ordinary skill *in the art to which said subject matter pertains.*" What good does it do some individual sweating in a cow barn shoveling out manure for there to be, far above, a spacecraft heading serenely toward the moon? John Reilly, of Adams fame, said:

> Now you know there's a very fundamental problem in Deere [Graham], Black Rock, and Ag Pro, and the fundamental problem is that there seems to be a sort of an intellectual superiority concerning the mundane arts.

---

*The Supreme Court opinion (as was quoted on page 69) said that the spray-cap invention "rests upon *exceedingly small* and quite *nontechnical mechanical* differences in a device which was old in the art." (italics mine)

> Deere concerned a plow; Ag Pro concerned cleaning a cow barn; and Black Rock was asphalt paving. I think you must persuade a court that our modern minds find it easy to see a patentable invention in matters relating to chemistry or electronics because what occurs in those fields is new to us. Because we are largely scientifically untrained, it is new territory to us and it is new territory to a judge. I disagree with the dictum in A&P that "elements may, of course, especially in chemistry or electronics, take on some new quality or function from being brought into concert, but this is not a usual result of uniting elements old in mechanics." . . . That is *scientific snobbery and I am certain it is not true.* (italics mine)

For some reason, most of the cases selected by the Supreme Court for review of patentability are nontechnological and "simple." The fact is, the vast majority of United States patents are issued for technological subject matter. When nontechnological patents are declared invalid, the resulting law is often applied to both technological and nontechnological subjects. Thus, the inventor of a new atomic reactor would, were he to arrive in court in pursuit of an infringer, be met by Supreme Court cases relating to the cow barn, road paver, supermarket counter, and so forth.

Perhaps, in the final analysis, it all comes down to a subconscious feeling on the part of judges and others that the technological and complex subjects are important to humanity, while the prosaic and simple ones are not. My point is that unobvious advances that happen to be simple and nontechnological are often very important and should not be discriminated against.

The Men

The monopoly stigma, the long-ago actions of greedy individuals in using patents in noxious ways, the lack of understanding of the patent system and of the nature of inventions—all tended to make many judges strict or hyperstrict. But those factors are not the final answer as to why patent owners have had rough times in court. The final answer lies in the men, the persons in high office during critical periods of our history. Who were they? And to what extent were they influenced by the stated factors?

In the Antitrust Division, there was Thurman Arnold, whose dialogue with Patent Commissioner Coe has already been quoted. He launched what has been called the first antitrust crusade relative to patents, and by winning case after case he stamped out many evils generated by the greedy men and their misuses of patents. (Later, men came into the Antitrust Division who did a great deal toward trying to eliminate other evils that, for the most part, did not exist.)

By far the most important men affecting the treatment of patents were on the Supreme Court—the mid-century Supreme Court. It set the tone for patent law for almost forty years (1939–1976). And, on that Court, two Justices largely controlled the course of patent law for the four decades.

The two were famous, great, well-intentioned Justices who were and are highly regarded, for example, relative to their work in fields such as civil rights. They wrote forcefully and articulately, having strong feelings and expressing them with enormous vigor. In patent cases, these feelings were almost always directed against the patent owner. In sum, these two Justices were influenced, to an extreme degree, by the factors discussed earlier in this chapter.

The two Justices were Hugo L. Black and William O. Douglas. They almost always voted together in patent cases, and it was common knowledge among lawyers for inventors that winning in the Supreme Court required the vote of five of the remaining seven Justices in almost every case. Five out of seven is tough to get and was seldom obtained.

There were no Justices on the Court as lenient toward

patents as Justices Black and Douglas were hyperstrict toward them. In fact, there was not then (nor has there been since) *any* Justice who may be classified as at all "lenient" relative to patents. Thus, the strongly held views of Justices Black and Douglas were to no extent balanced by equally strongly held (but lenient) views of other Justices. Justice Felix Frankfurter was too scholarly and restrained to write as strongly in favor of patents as Black and Douglas wrote against them. Justice Frankfurter did occasionally burst forth with an eloquent and powerful dissent, as when the main claims of the Marconi radio patent were invalidated, but he and other Justices who were not hyperstrict usually went with the flow and did not raise their voices.

Justice Douglas sat on the Court from 1939 to 1975, longer than any other Justice in history. Justice Black, joining the Court in 1937, was a member only two years less and actually voted in more test-of-patentability cases than Douglas—one reason being that a small flurry of such cases came before the Court between 1937 and 1939.

Justices Black and Douglas waged no less than a crusade to reform the patent system and the Patent Office, and to impose very strict tests of patentability. During Justice Black's first term, seven cases relating to validity or enforceability of patents came before the Court. Three were adjudged in favor of the patent owners, and four against. In every case decided in favor of the patent owner, Justice Black was the sole dissenter.

While Justices Black and Douglas, or either of them, were on the Court, the majority of Justices upheld validity only four times in test-of-patentability cases. Justice Black voted against validity twenty-two times and for validity twice (the Adams case, and a case relating to the welding flux used in welding, for example, World War II "liberty ships"). Both of the patents for which Justice Black voted were for chemical or electrochemical, not mechanical, subjects.

Justice Douglas voted against validity eighteen times and for validity twice, in the same two instances as Black.*

In the "liberty ship" case, the majority of the Court ruled that there was infringement; Justices Black and Douglas were the only dissenters. (There were two majority opinions in the "liberty ship" case, one relating to validity and one to infringement. Both were written by Justice Jackson, and both favored the patent owner relative to the flux claims. Other claims, for a method, were invalidated.)

The heart of the hyperstrict standards applied by Justices Black and Douglas involved their belief that a patent represents "a narrow exception to our competitive enterprise system." They were not even satisfied with the extremely antipatent opinion by Justice Jackson in the A&P case. Instead, Justice Douglas wrote, and Justice Black joined in, a celebrated "concurring" opinion. (A concurring opinion agrees with the vote of the majority, but for different reasons.) This concurring opinion achieved far more publicity than did the majority opinion of Justice Jackson. For example, the beginning (reproduced here) of *The New York Times* account related primarily to the concurring opinion.

---

*It is emphasized that the voting records stated in this paragraph, and in the preceding one, relate solely to test-of-patentability cases. (It is not, however, meant to be implied that the voting records of Justices Black and Douglas were more favorable to patents in other types of patent-related cases.) The expression "test-of-patentability cases" refers to those cases in which some test, such as obviousness and so forth, is applied to a particular patent. The expression does not refer to other types of cases, such as those in which invalidity is based on an inadequate disclosure, defective claims, or procedural matters. Also, the expression does not refer to cases in which the issue is the patentability of a *whole class* of subjects, such as a certain type of chemical. In a case relating to the latter, Douglas voted for the patent applicant while Black joined the majority and voted against. (My opinion that the patent decisions of the mid-century Supreme Court were largely controlled by Justices Black and Douglas is not, of course, based solely on the voting records of these two Justices. It is based, also, on numerous other factors, including: (1) what they wrote about patents in their opinions, (2) their long tenures on the Court, and (3) feelings expressed by knowledgeable members of the Patent Bar. For example, Reilly's argument in the Adams case was directed primarily at Justice Douglas because Reilly believed that the opinion of that Justice was pivotally important.)

HIGH COURT SHOWS SOME PATENT IRE

It Points to the Constitution, Bringing a Ruler Down on the Inventions Office's Knuckles

GADGETS CAUSE OF IT ALL

Law's Protection Doesn't Work for Petty Thingumbobs, the Tribunal Decrees

WASHINGTON, Dec. 4 (AP)—The Supreme Court lost its patience today with gadget-makers who wanted patents and it cracked the knuckles of the Patent Office for humoring them.

Henceforth, the Court decreed, the test for granting of patents must be stiffened and must hew closely to constitutional standards.

The framers of the Constitution, the Court said, "never sanctioned the patenting of gadgets. Patents serve a higher end—the advancement of science."

And, to back up its contention, the Court quoted Article One, Section Eight of the Constitution, which says the purpose of patents is "to promote the progress of science and useful arts."

The Patent Office, the Court said, "has placed a host of gadgets under the armor of patents" in seizing an opportunity to expand its own jurisdiction. In doing this, the Court added, the Patent Office has acted "like most administrative agencies."

2 Plus 2 Still Make 4

The high tribunal tossed out a contention of the Supermarket Equipment Corporation that the Great Atlantic and Pacific Tea Company had infringed three claims involved in a patent. The patent at issue covered a tray-like device for store check-out counters to pull customers' merchandise up to the checker.

The A. & P. appealed from a ruling of the United States District Court in Detroit. It contended the claims were invalid for lack of invention. The Supreme Court agreed unanimously.

Justice Robert H. Jackson said of the device in question:

"The three-sided rack will draw or push goods put within it from one place to another—just what any such rack would do on any smooth surface—and the guide rails keep it from falling off from the counter, as guide rails have ever done. Two and two have been added together and still they make only four."

Justice William O. Douglas put it more bitingly in a separate concurring opinion in which he took broad swings at the Patent Office. He wrote:

"The fact that a patent as flimsy and as spurious as this one has to be brought all the way to this court to be declared invalid dramatically illustrates how far our patent system frequently departs from the constitutional standards which are supposed to govern."

The beginning of the A&P account in *The New York Times*. ©1950 by The New York Times Company. Reprinted by permission.

In his opinion, Justice Douglas wrote:

> Every patent case involving validity presents a question which requires reference to a *standard written into the Constitution*. . . .
>
> Every patent is the grant of a privilege of exacting tolls from the public. The Framers plainly did not want those monopolies freely granted. The invention, to justify a patent, had to serve the ends of *science —to push back the frontiers of chemistry, physics, and the like; to make a distinctive contribution to scientific knowledge*. . . .
>
> The attempts through the years to get a broader, looser conception of patents than the Constitution contemplates have been persistent. The Patent Office, like most administrative agencies, has looked with favor on the opportunity which the exercise of discretion affords to expand its own jurisdiction. *And so it has placed a host of gadgets under the armour of patents —gadgets that obviously have had no place in the constitutional scheme of advancing scientific knowledge.* (italics mine)

The concurring opinion contains major inaccuracies. Contrary to what was said in that opinion: (a) the Constitution does not contain any "standard" of patentability, test of patentability, or whatever, relative to any *individual* invention; (b) there is nothing in the Constitution or patent law that militates against "gadget" patents or glorifies chemical or scientific patents; (c) it is well recognized by everyone— and by the Supreme Court in its Graham opinion —that the word "science" that appears in the Constitutional clause relates solely to copyrights, not to patents.* Patents are to promote the "useful arts."

---

*"The Congress shall have power . . . To promote the progress of science and useful arts, by securing for limited times to authors and inventors the exclusive right to their respective writings and discoveries." "Science" goes with "authors" and "writings," while "useful arts" goes with "inventors" and "discoveries."

Judge Rich discussed the concurring opinion in 1978:

> Justices Douglas and Black, . . . Douglas writing the *concurring* opinion, said that "The standard of patentability is a Constitutional standard . . ." and that "The question of invention goes back to the Constitutional standard in every case." He said the restraint on Congress was in the Constitution's statement of purpose. He emphasized his point by a tirade against "gadget" patents on clay doorknobs, pencils with erasers, hairnets, staples, cue racks, suspenders, corsets, pants flies, oval toilet paper rolls, and many other patented inventions which he called "incredible patents" which the Patent Office has "spawned" under what he assumed to be a bureaucratic desire "to expand its own jurisdiction." His idea of a constitutional standard, which he clearly equated with a proper standard of "invention," was that an invention, to justify a patent, must "serve the ends of science—to push back the frontiers of chemistry, physics, and the like; to make a distinctive contribution to scientific knowledge." If that . . . is not a gross misreading of the patent and copyright clause of the Constitution, I have never seen one. *He selected the statement of purpose pertaining to the writings of authors and totally ignored the statement pertaining to the discoveries of inventors and the promotion of useful arts.* (italics, except the word "concurring," mine)

Thomas Jefferson did not appear to know of any test of patentability written into the Constitution. He sought proper tests and proposed the obviousness test. Then, for decades after the three-man board ceased operating, patents were granted, and sustained in court, without any patentability tests. It wasn't until the doorknob case in 1850 that the first judge-made test came along.

In 1875, in the famous pencil-eraser case, the question was raised whether or not the Supreme Court had the power *at all* to invalidate a patent except on grounds specified by Congress. The writer of the opinion was not outraged by this suggestion, but looked solely at the patent act of Congress, and did not make a single reference to the Constitution. He said, "It is nowhere declared in the Statute that the decision of the commissioner, as to the extent of the utility or importance of the improvement, shall be conclusive upon that point. . . ." It may be inferred that if the statute had declared the commissioner's decision to be conclusive, that would have ended the matter.

Fortunately, the "constitutional standard" is ancient history—at least as related to individual inventions. Although the standard was stated in the Graham opinion, and again in the Black Rock opinion written by Justice Douglas, both opinions were referring to the power of Congress, under the Constitution, to establish patentability tests. (That matter is unrelated to application of any constitutional standard to particular patents on a case-by-case basis.)

Said the Supreme Court at the beginning of the Graham preface:

> Within the limits of the constitutional grant, the Congress may, of course, implement the stated purpose of the Framers by selecting the policy which in its judgment best effectuates the constitutional aim. . . . Within the scope established by the Constitution, Congress may set out conditions and tests for patentability.

Such "limits" and "scope", wrote the Court, were to be found in the Constitution's statement of purpose; thus, any test created by Congress must be such as to promote the progress of useful arts.

The patent legacies of Justices Black and Douglas are far reaching. They include, primarily, hyperstrictness and major additions to the confusion about patentability tests. Yet, neither Justice Black nor Justice Douglas was anti-inventor or anti-invention. Justice Douglas, in fact, wrote an important tax opinion that was liberal in its giving of income-tax incentives to persons who back inventors.

The hyperstrictness of both Justices concerned patent law and other types of law giving *exclusivity* to inventors. As will be described in the following chapter, Justice Black took the leading role (joined, but to a lesser extent, by Justice Douglas) in Court opinions severely adverse to exclusivity in nonpatent areas of inventor-related law. It was these opinions that—by a twist of fate—led to a major turning point in the Supreme Court, a turning point foreshadowing decisions and opinions favorable to several forms of inventor exclusivity.

TEN

Vastly Better Patent Pending

Laws and institutions must go hand in hand with the progress of the human mind.
THOMAS JEFFERSON

Imagine court decisions, and the climate for invention, as being analogous to a pendulum. From a position of relative leniency in the 1800s and early 1900s, the patent pendulum, reflecting court decisions, swung toward the hyperstrictness of the post-World War II period. But even as it approached the extremity of its arc, it was losing momentum, so that at the point where the judicial climate for inventors seemed at its worst, in 1976, the pendulum was actually beginning its swing back to a more favorable position.

During most of the antipatent period, the general public, along with the average inventor, had little or no awareness of how bad things were. Then, starting about 1980—when the patent system was improving dramatically—the mass media discovered what had been going on. (During the decades when many courts were tightening the screws on patents, the mass media did an inadequate job of reporting. Of course, such cases as A&P and AGPRO were written up, but the public was not informed that they were symptomatic of a general decline and rampant confusion.) A flurry of reports appeared, describing the more spectacular of the problems with the system.

One of the first exposés was printed in *The Washington Post* in the summer of 1980: "Patent Office Fails as Mother of Invention," read an August 18 *Post* headline.

" 'It was the damnedest thing I'd ever seen,' " the newspaper quoted a part-time St. Louis inventor, William P. Outten. " 'You feel like you're just in the attic of the United States.' "

Continuing its story, the *Post* wrote, "In two days, William Outten had learned what inventors, patent lawyers—and patent commissioners—have been worked up about for years: The Patent and Trademark Office is a mess. The office operates as though the high-powered technology described in its files didn't exist. Its 24 million documents are kept not on microfilm or computer, but on millions of pieces of paper scattered through hundreds of small rooms in three separate buildings."

Later in the article, the *Post* quoted from a speech by former Patent Commissioner Donald W. Banner: " 'We are faced with a slowly but steadily declining Patent and Trademark Office. . . . If we do not promptly reverse this direction of movement, it shall soon be infected with an administrative dry rot condition, rendering it moribund.' "

The most widely received exposé was by "NBC Magazine," which aired a program on December 5, 1980. After a statement by David Brinkley, the program's Douglas Kiker interviewed three inventors, all of whom had sued alleged infringers. Two lost their cases and one was still before a court. Those interviewed included George Ballas, inventor of the "Weed Eater" garden tool that whirls a plastic line sufficiently fast to cut weeds:

> *Kiker:* When you're invited for dinner at George Ballas's home, you eat well. Ballas is an entrepreneur, a real estate developer, who started off as a poor boy in rural Louisiana and ended up rich in Houston. Ballas is the inventor of the Weed Eater.
>
> *Ballas:* The idea came for a need and finding a way to trim the weeds growing round my trees and one day in a car wash I watched those brushes go round and I thought to myself by golly that's what I need. But I found some radio wires at first and just slipped them off and tried to simulate, you know, a car wash brush going into these trees and pulling out that turf.
>
> *Kiker:* So, associating with a machinist, Ballas perfected his Weed Eater, got it patented and started manufacturing and marketing it. It was an instant commercial success and Ballas thought he had it made . . . until more than 40 other companies quicky came out with their own versions.
>
> *Ballas:* When the first infringers started coming out, I ran to my attorney and said, hey, we got an infringer, let's stop him. He said, well, you can't stop him. Why not? Well, you see your patent isn't really valid until it's proven in court. In other words, you're guilty until you go to court and prove you're not guilty. That was a shock.
>
> *Kiker:* Ballas went to court and he lost. . . . One reason Ballas lost his case was because the judge ruled his invention was obvious. You cannot patent an invention that would be obvious to someone skilled in the subject. For example, you could make a coffee cup with a square

> handle and apply for a patent but you wouldn't get it because it's too obvious. The fact is some judges believe that patents create monopolies, they are against monopolies so they are against patents and they use this doctrine of obviousness to invalidate patents.

Most of the revelations by the mass media were generally deficient in several respects:

1. Being overly negative.
2. Directing an excessive proportion of the blame toward the Patent Office.
3. Failing to see and report, to a sufficient degree, that things are turning around in the courts, especially in the Supreme Court.
4. Not reporting adequately the patent-policy reversals in the Antitrust Division, which largely determines what people can do with their patents without being hassled.
5. Not emphasizing at least one monumental pro-patent occurrence in Congress, which not only generates the patent laws but has the crucial power to determine what courts shall administer them.

The Turnaround in the Courts

Fewer than 1 percent of patents are ever brought to trial. During the ordinary, prosaic course of events, applications are filed, inventions are patented, and products are put on the market. When infringements occur, the parties often negotiate agreements resulting in licensing or the cessation of infringement. Court action is the last resort. Then why talk so much about court cases? Because the court system acts as the primary regulator of the whole patent mechanism: it is the judges who interpret the laws and establish the precedents under which the system operates, and who expand and contract its limits. It is they who decree what can and cannot be done, and who, by their rulings, affect the climate for invention negatively or positively. It is they who affect the value of *all* patents, especially during the long periods between major acts of Congress.*

In order to provide background information for the turnaround in the courts, an important concept generated by the mid-century Supreme Court, that of preemption, will be mentioned. Preemption is the idea that when Congress created the patent system, other forms of protection for the inventor were squeezed out and obliterated—"preempted." Carried to the extreme, this interpretation would mean that if an inventor did not have a valid patent, he would have no protection. State law prohibiting the stealing of trade secrets and providing for the enforcement of certain types of contracts would be emasculated.

The theory of preemption was first stated by Justice Black, in opinions delivered on March 9, 1964, over a century and a half after the patent system was established. On that date, the Supreme Court decided two separate cases involving the law of unfair competition. The first case, known as "Sears," dealt with an enormously successful pole lamp that Sears, Roebuck & Co. had copied from another firm. The second case, called "Compco," pertained, in similar fashion, to a fluorescent lighting fixture. Since neither the pole lamp nor the fluorescent fixture was validly patentable, Sears and Compco were sued primarily on the ground that their products "confused" purchasers, who might have thought they were buying the original product rather than a copy.

The Supreme Court decisions in the two cases were that state law relating to unfair competition could not prevent the copying of unpatentable products. These decisions may have been correct under the particular fact situations that were present, but as often happened, the language of the opinions far exceeded the decisions, and staggered the patent

*For example, had the court decisions been different, the letter from the patent lawyer to Vanity Fair (summarized on pages 16 and 17) would probably have been different, and Vanity Fair would not have proceeded. Olga's patents might then have remained unlitigated, but—unbeknownst to her—would have been valuable indeed.

bar. In his usual sweeping terminology, Justice Black said: "Just as a State cannot encroach upon the federal patent laws directly, it cannot, under *some other law,* such as that forbidding unfair competition, give protection of a kind that clashes with the objectives of the federal patent laws." (italics mine)

Other language used by Justice Black made it clear that if a product was not covered by a *valid* patent, it was not protected at all. In other words, the preemption concept involved two facets: (1) whether there was a patent, and (2) whether the patent was valid.

Five years after the Sears and Compco cases, the Supreme Court used preemption as a major reason for taking away a basic protection that the law of contracts had long given to patent owners who grant licenses to others. In the landmark "Lear" decision, the Court decided that a licensee could attack the validity of the licensed patent even while the license was still in effect—while he was still enjoying its benefits. If the licensee won his fight, he could use the invention free. And if the licensee lost, he did not have to stop using the invention—instead he merely made royalty payments just as if there had been no fight. (All the while, if the license was exclusive, the patent owner could not license another or even use his own invention.)

The rationale behind the Lear decision was that licensees would not be inclined to try to wipe out patents unless given this advantage (they would, said the Court, be "muzzled"). And, if they (as persons having the most incentive) didn't try to wipe out the patents, some *invalid* patents might be earning royalties. This would be contrary to the Sears-Compco concept that only valid patents should be effective.

The majority of Justices in the Lear case did not decide what happens to licenses under patent applications as distinguished from issued patents. (That question was sent back to a lower court.) But some Justices, including Justice Black, who was joined by Chief Justice Warren as well as Justice Douglas, wanted the Supreme Court decision to apply to licenses under patent applications as well. Patent applications are not available to the public, but are instead secret, so the Black-Warren-Douglas opinion brought into question the law of trade secrets. In a famous passage, Justice Black stated:

> I still entertain the belief I expressed for the Court in Stiffel [Sears] and Compco that no State has a right to authorize any kind of monopoly on what is claimed to be a new invention, except when a patent has been obtained from the Patent Office under the exacting standards of the patent laws. One who makes a discovery may, of course, keep it secret if he wishes, but *private arrangements under which self-styled "inventors" do not keep their discoveries secret, but rather disclose them, in return for contractual payments, run counter to the plan of our patent laws.* . . . The national policy expressed in the patent laws, favoring free competition and narrowly limiting monopoly, cannot be frustrated by private agreements among individuals, with or without the approval of the State. (italics mine)

Naturally, Justice Black's pronouncement encouraged assaults on trade secret law. Although that law (by which it is possible to sue those who take secret information and use it for their own benefit) had been repeatedly applied in federal and state courts, the Court of Appeals for the Sixth Circuit, in May 1973, felt obliged to follow the new path of preemption. In the case before the Court, Kewanee Oil Company v. Bicron Corporation, the incorporators of the latter company were accused of having applied secret knowledge, that they had gained as employees of Kewanee, in order to manufacture synthetic crystals.

The issue, the Court said, was "whether the trade secrets laws of the State of Ohio are required to be struck down in this case as being in conflict with the United States Patent Laws." Basing their decision primarily on Sears, Compco, and Lear, the judges stated that they were forced "to the conclusion that the field of protection afforded to this plaintiff by that

Trade Secret Law has been preempted by the Patent Laws of the United States."* Kewanee, therefore, lost in that Court.

Many of the patent bar were put into a state of near panic by the decision. Neither patents, contractual obligations, nor trade secrets seemed to offer sufficient protection any more. After the case was appealed to the Supreme Court, interest was so intense that no fewer than twenty-one amicus curiae (friend of the court) briefs were filed. There was great apprehension that the Supreme Court would continue the trend of recent years and uphold the Court of Appeals.

When Chief Justice Warren Burger, summarizing the opinion in May 1974, stated, "We hold that Ohio's law of trade secrets is not pre-empted by the patent laws of the United States, and, accordingly, we reverse," there was a sigh of relief among spectators in the Court chamber. Explaining the majority's reasoning, the Chief Justice had written:

> Certainly the patent policy of encouraging invention is not disturbed by the existence of another form of incentive to invention. In this respect the two systems are not and never would be in conflict. . . . Trade secret law and patent law have co-existed in this country for over one hundred years. Each has its particular role to play, and the operation of one does not take away from the need for the other. Trade secret law encourages the development and exploitation of those items of lesser or different invention than might be accorded protection under the patent laws, but which items still have an important part to play in the technological and scientific advancement of the Nation.

Justice Marshall concurred only in the result, while Justice Powell did not participate. Justices Douglas and Brennan

* The opinion emphasized, however, that a trade secret *could* be protected during the time period when it was legally possible to apply for a corresponding patent.

dissented, with Douglas writing: "Today's decision is at war with the philosophy of Sears . . . and . . . Compco," as indeed it was. (Justice Black was no longer on the Court.)

This decision was the first sign that there would later (after AGPRO!) come a shift, by a Supreme Court majority, away from the hyperstrict doctrines of the previous three decades.

What Kewanee did for upholding trade secrets, a subsequent Supreme Court case, called "Quick Point," did for upholding certain kinds of contractual obligations. In 1956, Jane Aronson, inventor of the widely used plastic key holder shown in the drawing, negotiated a contract with the Quick Point Pencil Company for the manufacture and sale of the

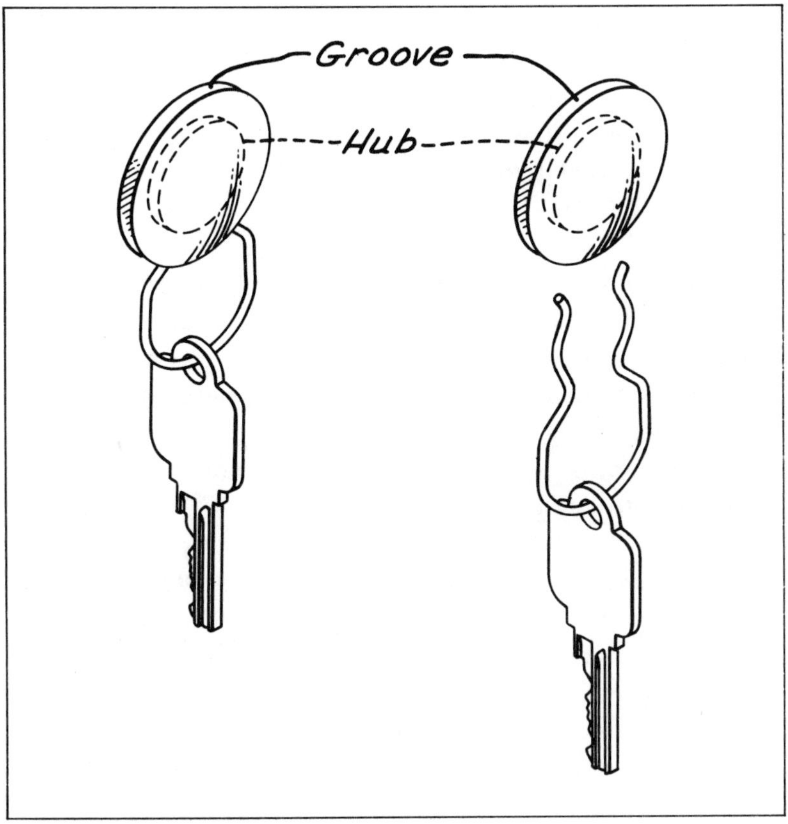

Aronson's plastic key holder.

device. The contract stipulated a royalty of 5 percent if Aronson obtained a patent within five years, and a royalty of 2½ percent, forever, if no patent issued. The Patent Office rejected the application, and the 2½ percent was duly paid until September 1975. (Total royalties paid up to that date were $203,000.) Then Quick Point, realizing that if a patent had issued it would by this time be expiring and the company would be freed from further royalties, commenced an action to terminate the contract.

In December 1977, the Court of Appeals for the Eighth Circuit, dividing two-to-one and citing Lear, Sears, and Compco, held, "Federal law requires that all ideas in general circulation be dedicated to the common good unless they are protected by a valid patent. . . . We believe this principle applies to a state's contract law as well as to state unfair competition laws. . . ." Thus, Jane Aronson lost in that court.

In February 1979, the Supreme Court, overruling the Court of Appeals and deciding in favor of Aronson, held unanimously that federal patent law did not preempt state contract law in the Quick Point situation:

> Enforcement of the agreement does not withdraw any idea from the public domain . . . [nor] discourage anyone from seeking a patent. . . . No decision of this Court relating to patents justifies relieving Quick Point of its contract obligations. . . . Enforcement of this royalty agreement is even less offensive to federal patent policies than state law protecting trade secrets.

State law, the Supreme Court said, was not superseded merely because a commercial agreement related to property that might or might not be patentable. It should be pointed out, however, that the Quick Point case did not overrule the Lear case, but instead limited its field of application.

The key holder of Quick Point represented the kind of simple subject matter at which Justices Black and Douglas had aimed their "gadget" blasts. But in "Chakrabarty," the Supreme Court plunged into the esoteric field of genetic engineering and produced one of the most liberal and progressive patent decisions in decades.

In 1972, Ananda Chakrabarty, a microbiologist employed by General Electric, filed a patent application that included claims for a new bacterium. The bacterium was claimed to have the ability to break down and degrade components of crude oil, and so to clean up oil spills. In effect, it "ate" camphor and octane, two of the components of petroleum. The Patent Office held that the bacterium was not patentable, ruling that living things are not covered by the following part of the Patent Act of 1952:

> *Section 101:* Whoever invents or discovers any new and useful process, machine, manufacture, or composition of matter, or any new and useful improvement thereof, may obtain a patent therefor, subject to the conditions and requirements of this title.

The section is not a new one; the key words date back to the Patent Act of 1793 and have been in the patent law ever since.

By a divided vote, the Court of Customs and Patent Appeals reversed the Patent Office. The majority opinion, written by Judge Rich, said, "This court *unanimously* believes it is not necessary that Congress shall have *foreseen* a new field of technology or useful art to bring it within § 101."

The stage was thus set for the Supreme Court, which decided the case on June 16, 1980. Only two years before, in 1978, the Court had indicated that it would be cautious and adopt a strict constructionist attitude, stating: "We must proceed cautiously when we are asked to extend patent rights into areas wholly unforeseen by Congress." It appeared, therefore, that the Court would uphold the government's arguments that "the terms 'manufacture' or 'composition of matter' [in Section 101] do not include living things [and] that micro-organisms cannot qualify as patentable subject matter until Congress expressly authorizes such protection."

Chief Justice Burger, in delivering the majority opinion, rejected these arguments. Quoting Thomas Jefferson that "ingenuity should receive a liberal encouragement," and pointing out that the Committee Reports accompanying the 1952 Patent Act indicated that Congress intended to "include anything under the sun that is made by man," the Justices voted five-to-four (Justices Brennan, White, Marshall, and Powell dissenting) squarely in favor of innovation. Wrote Chief Justice Burger:

> A new mineral discovered in the earth or a new plant found in the wild is not patentable subject matter . . . nor could Newton have patented the law of gravity. Such discoveries are "manifestations of . . . nature, free to all men and reserved exclusively to none." . . . Here, by contrast, the patentee has produced a new bacterium with markedly different characteristics from any found in nature and one having the potential for significant utility. His discovery is not nature's handiwork, but his own. . . .
>
> Congress . . . recognized that the relevant distinction was not between living and inanimate things, but between products of nature, whether living or not, and human-made inventions. Here, [the] micro-organism is the result of human ingenuity and research.

The bacterium was, therefore, held to be patentable subject matter under Section 101.*

Not only was the Chakrabarty decision of immense precedential importance, extending patent protection to an entirely new class of subject matter, but it was one of the most pro-patent decisions issued in recent decades by the nation's

---

*Chief Justice Burger made it clear that the Court was *not* deciding whether or not genetic engineering is good for society. That, he said, is up to Congress, which has the power to deny such patents if it wishes.

highest judicial body. Lawyers were still trying to analyze its significance when, only eleven days later, the Court provided clearcut affirmation—although by the same narrow five-to-four margin—that it would support enterprise and creativity. The case, Dawson Chemical Co. v. Rohm and Haas Co., involved the use of an unpatented product in a patented process. In the Rohm and Haas case, the unpatented product was named propanil. The patented process (patented by Rohm and Haas) was to apply propanil to rice fields to kill weeds without killing the rice.

There have for many decades been two important judge-made doctrines. One, *misuse of patents,* prevents patent owners from unconscionably extending patent coverage to unpatented goods. The second doctrine, called *contributory infringement,* is directed at persons who do not themselves infringe any patent but instead sell, to infringers, things that the infringers need in order to do their infringing. (There is no liability for contributory infringement unless there is knowledge of the particular patent being infringed.)

By a case decided in 1944, the Supreme Court—in an opinion by Justice Douglas relating to patent misuse—all but wiped out contributory infringement. The following statement is especially famous: "The result of this decision . . . is to limit substantially the doctrine of contributory infringement. What residuum may be left we need not stop to consider." As a result, Congress put contributory-infringement and patent-misuse sections into the Patent Act of 1952.

The sections represent a compromise between the two doctrines. They provide, essentially, that there can be contributory infringement, and no patent misuse, only in relatively unusual situations—namely, where the *sole* substantial use of the thing supplied is in the patented combination or method. (In the Rohm and Haas case, the sole substantial use of propanil was to kill weeds in rice fields when applied in accordance with the method patented by Rohm and Haas.)

The situation was similar to that regarding tests of patentability: uncertain court decisions not related to any statute, followed by a Congressional attempt at clarification by enacting a new law (statute). Would the attempt be fully successful this time? The answer, unlike that provided by the Graham opinion concerning the obviousness section, was yes.

On June 27, 1980, the five-Justice majority ruled clearly, definitely, and entirely in favor of Rohm and Haas. The majority opinion, by Justice Blackmun, is notable for several reasons, one being that it included this strong pro-inventor statement:

> The policy of free competition runs deep in our law. It underlies both the doctrine of patent misuse and the general principle that the boundary of a patent monopoly is to be limited by the literal scope of the patent claims. *But the policy of stimulating invention that underlies the entire patent system runs no less deep.* And the doctrine of contributory infringement . . . can be of crucial importance in ensuring that the endeavors and investments of the inventor do not go unrewarded. (italics mine)

In a story appearing shortly after the Rohm and Haas decision, a writer for *The Wall Street Journal* declared that the "new Supreme Court ruling placed the overlapping demands of patent protection and competition on the scales of justice and tipped the balance in favor of patents. Shedding new light on a controversy that has plagued patent law for most of this century, the Supreme Court ruled that a company that obtains a patent for the only known use of a chemical can also control the market for actual manufacture of the chemical, even when the chemical is unpatented. The ruling is important for at least two reasons. First, it may have a significant impact in the drug, pesticide and herbicide industries, where it isn't unusual for scientists to find new uses for old substances or chemicals that aren't used any longer. Second, it is another in a series of cases that suggest that the current Supreme Court advocates strong patent protection for innovative companies."*

On the other hand, the decision shocked some of the Justices. Justice Stevens, who had been a member of the majority in the Chakrabarty case, joined Justices White, Brennan, and Marshall in dissent. (Justice Powell switched positions with Stevens to provide the majority in Rohm and Haas.) Justice Stevens wrote: "This is a classic case of patent misuse. . . . The patent may well have little or no commercial value unless the patentee is permitted to engage in patent misuse. But surely this is not a good reason . . . to permit such misuse."

The most recent decision (as of early 1983) evidencing the turnaround in the Supreme Court was in the case of Diamond, Commissioner of Patents and Trademarks v. Diehr and Lutton. Diehr and Lutton had devised a process for molding raw, uncured synthetic rubber into cured precision products. Prior to the Diehr process, the industry had not been able to obtain uniformly accurate cures because of the inability of the operator to respond precisely to the various temperatures within the molding press. The Diehr and Lutton improvement on the standard operation involved the continuous measurement of the temperatures inside the mold and the automatic feeding of these measurements into a digital computer that recalculated the mold time and signaled the operator when to open the press.

The Patent Office rejected the claims. The only novelty, it said, was the control of the operation by computer, but computer programs had been held to be unpatentable as not being within Section 101 (quoted on page 195, and being the same section involved in the microorganism decision). The Court of Customs and Patent Appeals then reversed the Patent Office and held that the patent should issue.

On March 3, 1981, in another five-to-four decision (Jus-

---

*From *The Wall Street Journal*, reprinted by permission. © Dow Jones & Company, Inc. 1980. All rights reserved.

tices Brennan, Stevens, Marshall, and Blackmun dissenting), the Supreme Court upheld the Court of Customs and Patent Appeals. The opinion was written by Justice Rehnquist, and noted that "a process has historically enjoyed patent protection because it was considered a form of 'art' as that term was used in the 1793 Act. . . . Our conclusion . . . is not altered by the fact that in several steps of the process a mathematical equation and a programmed digital computer are used."

The majority opinion said, in effect, that although computer programs in the abstract were not patentable, they could be parts of patentable methods.* Of possibly greater significance, however, was one of Justice Rehnquist's statements in support of the decision:

> In determining the eligibility of [the] claimed process for patent protection under §101, [the] claims must be considered as a whole. It is inappropriate to dissect the claims into old and new elements and then to ignore the presence of the old elements in the analysis.

This statement gives rise to hope that the majority might in the future look at combination patents in a new light, even in test-of-patentability cases (obviousness Section 103, as distinguished from Section 101), returning to the 1882 Supreme Court philosophy that inventors "who make some substantial discovery or invention, which adds to our knowledge and makes a step in advance in the useful arts . . . are worthy of all favor."

What emerges from these crucial cases is a Supreme Court far less doctrinaire and more open-minded than its immediate predecessors. Each of the three most recent decisions (those relating to microorganisms, contributory infringement, and computers) was rendered by a *different* five-

---

*In a companion computer-related case that was not for a process, a decision by the Court of Customs and Patent Appeals—favorable to the patent applicant—was affirmed by an equally divided Supreme Court.

to-four majority—and each such majority contained Justice Potter Stewart. Justice Stewart has since resigned and been replaced by Justice Sandra Day O'Connor. Her views on patents are unknown but will be pivotally important, as will the views of other new Justices sure to be appointed in the next several years.

The Lower Courts

For decades the Supreme Court's negative attitude toward patents was transmitted downward. Lower-court judges sometimes uttered "slams" at patents, but positive comments on the patent system were so rare and ineffective as to be like voices crying in the wilderness. Recently, however, there has been a trend toward a different "tone" in many of the lower-court opinions, and a more favorable view of creative endeavor. This is not to imply that the opinions became uniform or became purged of the doctrines generated by the A & P case and its progeny. What is meant is that there shone forth in lower-court opinions a surprising number of clear and positive statements supporting the patent system.

For example, in a 1980 appeal involving an antibiotic that had taken Eli Lilly twenty years and $10 million to develop, the court stated that although companies which do not engage in significant research "might be able to undercut the prices offered by pharmaceutical manufacturers that devote large sums to invention and product improvement, this type of short-term competition does not, at least in the considered opinion of the Congress, serve the public interest. Instead, Congress has determined that it is better for the nation in the long-run to afford the inventors of novel, useful, and nonobvious products short-term monopolies on such products than it is to permit free competition in such goods."

Another definitive statement that a patent represents a legal, beneficial "monopoly" as long as it is not misused was made by the Court of Appeals for the Second Circuit in

SCM Corporation v. Xerox Corporation, decided March 12, 1981. To appreciate the full significance of this decision, I will describe what the *government* had already done to Xerox.

In January, 1973—only about thirteen years after Xerox Corporation was first able to put an automatic copier on the market—the Federal Trade Commission (FTC) filed an antitrust complaint against it. Xerox later settled, by accepting in 1975 a consent order, instead of expending its resources on a protracted and wasteful battle such as was recently finished by IBM. "The central purpose of the consent order," said the FTC, "is to eliminate the fundamental sources of Xerox's total dominance of the multi-billion dollar office copier industry—its vast patent portfolio, its manufacturing know-how and, to a lesser degree [a particular pricing program for its machines]." The consent order went a long way toward destroying Xerox Corporation's patent position.

As often occurs when the government brings an antitrust action, the FTC action stimulated a *private* antitrust claim, the purpose being to obtain mandatory treble damages. After Xerox refused to license SCM in 1964 and 1969 (long before the consent order), SCM—in July, 1973—charged Xerox with an illegal monopoly of the plain-paper copier market. Following a trial (and jury deliberations) lasting 257 days, a jury awarded SCM $37.1 million in damages, which were automatically trebled under the antitrust laws to $111.3 million.

The trial judge, and then the appeals court, overturned this generous award. Said the appeals court in its 1981 opinion:

> The law is unsettled concerning the effect under the antitrust laws, if any, that the evolution of a patent monopoly into an economic monopoly might have upon a patent holder's right to exercise the exclusionary power ordinarily inherent in a patent. . . .
>
> SCM has contended that a unilateral refusal to license a patent should be treated like any other refusal to deal by

> a monopolist. . . . [but] no court has ever held that the antitrust laws require a patent holder to forfeit the exclusionary power inherent in his patent the instant his patent monopoly affords him monopoly power over a relevant product market. . . . If the threat of treble damage liability for refusing to license were imbedded in the minds of potential patent holders as a likely prospect incident to every successful commercial exploitation of a patented invention, the efficacy of the economic incentives afforded by our patent system might be severely diminished. . . . We hold that where a patent has been lawfully acquired, subsequent conduct permissible under the patent laws cannot trigger any liability under the antitrust laws.

Therefore, unlike the FTC, the courts did not penalize Xerox Corporation for achieving its patent-based market dominance.

The Turnaround in the Antitrust Division

If Chester Carlson had had the foresight (and longevity) to wait about eight years before inventing xerography, thus pushing back the entire Xerox-related time scale, it is highly likely that the FTC would not have pressured Xerox Corporation into giving up the bulk of its patent position. This is because the FTC has been notably quiet of late, at least in the field of patents. Of much greater importance, however, is the reversal that has occurred in the powerful Antitrust Division of the Department of Justice.

The first (anti)patent crusade, launched by Antitrust Division Chief Thurman Arnold in the 1930s, had made sense, though he sometimes went overboard. But if Arnold went overboard in a few matters, men who conducted a second patent crusade during the 1960s and early 1970s went off the deep end. Traveling about the country, they harped on the theme of restricting the exercise of patent rights, attacked one patent licensing practice after another, and successfully fought attempts to obtain legislation making certain licensing practices legal per se. Eventually, in the rock-bottom spring of 1976, came their success in pushing through the Senate an entire new patent law (patent code), written by the Antitrust Division and embodying many of the Division's anti-patent prejudices. (The bill did not pass in the House; as mentioned on page 148, it never became law).

In 1973, a prominent antitrust lawyer, Earl Pollock, remarked:

> Once upon a time, patent law and antitrust law were two neighboring kingdoms, [and] for the most part they lived in harmony. . . . As we know, those halcyon days are gone forever. The aggressor, of course, has been antitrust, which has shamelessly asserted its jurisdiction over all forms of intellectual property. . . .

That same year, a prominent patent lawyer, Tom Arnold, complained: "I think it is terrible that our licensing practice is governed by what I call 'Department of Justice luncheon speech law.' . . . The Justice Department no longer feels bound by the case law [judge-made precedents] to any extent at all." The luncheon speech law emphasized, among other things, what were called "the nine no-no's of patent licensing."

The patent-related actions by the Antitrust Division slowed greatly under President Carter, when Deputy Assistant Attorney General Ky P. Ewing, Jr. made some speeches favorable to innovators. Then, in November of 1981, Ewing's successor—Abbott B. Lipsky, Jr.—came forth with a highly pro-patent speech, rejecting the contention "that the nine no-no's have much independent validity as economically rational antitrust rules. . . . When one makes the analysis, one finds that the nine no-no's, as statements of rational economic policy, contain more error than accuracy." In sum, Lipsky said that while the nine no-no's might have validity in individual cases, "[I] have found in almost every instance that they are overinclusive or contain at least some element of economic irrationality."

There are many other ways in which the Antitrust Division has shown that it is no longer against patents and other forms of inventor protection. For example, in 1978 the Division noted approvingly that the government had filed a brief in *favor* of Jane Aronson in the Quick Point case.

The Tremendous Turnaround in Congress

No more does the Antitrust Division try to enact patent legislation. Furthermore, the Division does not now take anti-patent stances concerning patent bills introduced by others. For this and other reasons, almost every patent-related bill passed by Congress after 1976 has been favorable to inventors and patent owners! Only a few of the most important bills are described here.

Since 1891, the greatest need of the patent system has been to have a *single* appellate court that would apply—nationwide—a *single* test of patent validity.* In 1974, Thomas E. Kauper, then head of the Antitrust Division, strongly opposed the concept, saying: "The search for uniformity by creating a single patent court of appeals seems simplistic and artificial. Moreover, it does not seem necessary, at least any more necessary in patent cases than any other category of federal question cases." But in (and after) 1978, the Division did not oppose a proposal made by Daniel J. Meador, Assistant Attorney General for Improvements in the Administration of Justice (a part of the Department of Justice). Meador suggested that the Court of Claims and the Court of Customs and Patent Appeals be merged to create a new federal appellate court. Under the proposal, the new court would handle everything then handled by the two separate courts, plus all

*Because it does not have time to hear more than a few patent cases, the Supreme Court cannot fill this need.

patent, civil tax, and environmental appeals. It was desirable, Meador said, to "resolve the myriad evils caused by fragmented review in tax, patent, and environmental litigation. The rampant lack of uniformity between the Tax Court, the district courts, the Court of Claims, and the regional courts of appeals would be cured. The forum-shopping common to all three areas of litigation would be cured. Business planning would be made easier as more stable law is introduced in all three critical areas."*

After introduction in March 1979 of Senate bills based generally on Meador's proposal, there followed more than three years of frustration on the part of the great majority of those who wanted a markedly improved patent system. The frustration and delay were not caused by any absence of support. The Carter Administration, and later the Reagan Administration, were strongly behind the original and subsequent bills in both Senate and House. (One assistant attorney general even testified at a House subcommittee hearing that the disparity in patent validity standards throughout the country is "grotesque.") Nor was there a lack of votes. Every time any single-court bill was put to the test, it won by an overwhelming margin. There was some opposition to the bills (mostly by the Section of Litigation, American Bar Association; the primary ground of resistance being that *any* specialized courts are not desired), but that opposition was a distinct minority and not at all effective.†

The three-year vigil that single-court supporters went through was occasioned not by opponents but largely by two well-meaning Senators, neither of whom opposed the single court!

---

*"Forum shopping" is trying to get a case tried in the place where you are most likely to win. Patent validity, as we have seen, varied enormously from circuit to circuit.

†The Patent, Trademark and Copyright Section of the American Bar Association favored the single-court concept, but the official ABA position was that of the Section of Litigation.

It is an unfortunate but common legislative tactic to attach "nongermane" amendments (riders) to strongly supported bills. In 1980, Senator Bumpers of Arkansas single-handedly shot down a single-court bill that was otherwise certain to become law. He did this because, though he was not against the single court, he was adamantly in favor of his amendment, which had nothing whatever to do with patents or inventors.

Then in late 1981, the floor manager and strong proponent of a single-court bill, Senator Dole of Kansas, surprised many by adding to that bill his own nongermane amendment. Happily, the bill eventually passed, as Senator Dole had anticipated it would.

President Reagan signed the single-court bill into law on April 2, 1982, with its date of effect being October 1, 1982. (As passed, the law does not include the civil tax or environmental matters that Meador had proposed.) On October 1, appeals from the federal district courts in patent cases started going to the United States Court of Appeals for the Federal Circuit. That Court is the first new federal court in over seventy years, and it is the very first appellate court, excepting only the Supreme Court, having nationwide jurisdiction.

It is almost certain that the new Court will—unless the Supreme Court directs otherwise—apply the Section 103 obviousness test as the sole one. The word "synergism" will not be uttered in nonchemical cases. No different test will be applied to "combination" inventions, nor will there be discrimination against mechanical subjects. Signpost evidence will be considered in all instances where it is present. A patent once held valid or invalid in that Court will be valid or invalid nationwide, except on those occasions when a new defendant finds undiscovered prior art that is more relevant than what was previously known.

Without doubt, the beneficial effects produced by the single-court bill will transcend those produced by anything that has happened to the patent system since creation of the Patent Office in 1836.

Other bills focused, very properly, on the Patent Office. In December of 1980, President Carter signed a bill directing that plans be made to achieve computerization in that venerable institution. To provide the necessary funds, the same bill mandated that Patent Office fees be increased, including "maintenance fees" due three and one-half, seven and one-half, and eleven and one-half years after a patent issues. Nonpayment of any maintenance fee causes the patent to lapse (there is a grace period).

The new fees were regarded by Commissioner of Patents and Trademarks Gerald J. Mossinghoff as being insufficiently high to really support modernization of the Patent Office or to maintain an adequate number of examiners. So, Congress passed another bill that raised the fees still further, with almost all of the increase to be paid by the larger corporations. This latest bill has some bright silver linings unrelated to fees. Most importantly, the bill guarantees that opposing parties who want to arbitrate a patent battle—including the crucial validity and infringement issues—can do so *bindingly*, so that the loser cannot drag the controversy into court. This has, strange to say, not been possible before.

In a speech delivered in August, 1982, Commissioner Mossinghoff said euphorically: "Everything that has happened in the world of patents . . . this past year has reinforced my optimism. . . . We have taken several important steps to move realistically toward a fully automated Patent and Trademark Office in the 1990s. . . . We have already installed computer terminals in each of our 15 patent examining groups to give patent examiners access to all available commercial patent data bases."

During the same speech, the Commissioner commented on recent legislation, emphasizing, among other things, (1) that arbitration authority "will provide business executives with an attractive and cost-effective alternative to protracted and complex Federal litigation in patent cases," and (2) that the new "single authoritative tribunal to handle cases nationwide . . . will contribute greatly to a single standard of patentability which will be understandable to inventors and business executives alike."

Recommendations for Improving the Patent System

There clearly has been great progress in the patent system since the dark days of the mid-1970s and before. Nevertheless, a number of key problems still remain. Additional action is required in three major areas covered by this book: Congress, the Patent Office, and the courts. Because Patent Office modernization is now under way, the following recommendations relate solely to Congress and the courts.

Suggested Statutory Changes Regarding Validity

The all-important obviousness Section 103 of the 1952 Patent Act should be amended by Congress to provide a *single, uniform, statutory* test of patentability. Whether or not the subject matter is a combination of old elements should be stated to be irrelevant. The sole issue, concerning all types of inventions, should be whether the subject matter as a whole would have been obvious to a person having ordinary skill in the art. (The type and size of differences between the prior art and the subject matter sought to be patented should—by the wording of the amended section—not be controlling; a crucial thing should be the importance of the differences.)

The use of catchwords should be negated by statute, whether they are used as supposed tests *or* as supposed indications of obviousness.

The section should specify that the test of patentability is to be applied reasonably (not strictly or leniently); and in a balanced manner to all subject matter, be it chemical, electrical, mechanical or biological, technological or nontechnological, complex or simple. It must be made clear that simplicity is not a negative characteristic, because patentable invention is frequently a process of progressing from the complex to the simple.

Signposts—for example, long-felt need for the invention, commercial success, and opinion of experts at the time of the invention—should be elevated, by statute, from secondary to primary status. Evidence on these points should be admitted and given serious consideration in all cases where such evidence exists.

Another badly needed reform is to not penalize research within corporations. For example, in a complicated case involving General Motors and Toyota, a district judge ruled (in 1979) that, although Toyota had copied GM's catalytic converter muffler, Toyota was not guilty of infringement because the patent was invalid. A major basis for the invalidation was that General Motors' patent resembled *its own* earlier, unsuccessful devices. In other words, GM's own research was held against it.* That this was an incredible interpretation of the patent law, and that the judge (who had followed some precedents) was subsequently overruled by a court of appeals—which ordered that a new trial occur—does not remove the necessity for preventing, by statute, a repetition of such a decision.

The purpose in suggesting these statutory changes relative to patent validity is not to initiate an era of hyperleniency or leniency, which could be as bad as hyperstrictness and strictness. What is wanted are reasonableness, understanding, uniformity, predictability, and balance. Of course, these goals could only be achieved if God himself were the judge; since judges are human there will always be some subjectivity, error, and occasional unfairness. However, the fact that perfection is not achievable should not prevent doing whatever is possible.

Any changes in the law, and the "Reviser's Notes" that accompany them, must be such that the standard of patenta-

*Of course, if any company's (or individual's) earlier efforts have been publicly used—nonexperimentally—for longer than is permitted by law, that *is* prior art. What I am talking about here are experimental, and usually secret, earlier efforts.

bility in the Patent Office is *not* lowered. This should not be hard, especially since the obviousness test (Section 103) is the only one used by the Patent Office.

Because the new Court of Appeals for the Federal Circuit will—in all probability—use obviousness Section 103 as the only test and will do many of the things suggested here, why are statutory changes needed? Why must Congress, to which the Constitution gave the "power" to establish the laws, exercise that power again—at least in the test-of-patentability area? Because: (1) Supreme Court decisions change with the passing decades, due to such factors as new appointments and new philosophies; (2) the new Court is directly under and easily controlled by the Supreme Court; and (3) even the policies of the new Court are uncertain and may change in the future (the Court of Customs and Patent Appeals was once distinctly strict). *Congress* must establish a correct standard of patentability that will continue far into the next century.

Need for Additional Patent-Experienced Judges

A substantial number of federal district judges have—through desire, experience, and study—become proficient in handling patent cases. Accordingly, any district judge now in office who *wants* to preside over patent cases should certainly do so. Many district judges, however, have no desire for patent cases and regard them as burdensome, to say the least.

Patent cases are often complex, time consuming, and difficult to handle. In April 1980, District Judge Frank J. McGarr remarked: "In the main they tend to be the most complex trials that we have. . . . It was lore in our district for many years that the way to punish a new judge for the nerve he displayed in coming on the court at all was to assign him all patent cases when he came on, and it took him five years to work his way through this burden. We finally straightened that out with a fairer assignment system, but it doesn't speak very well of the patent cases to reflect that those were regarded as a form of punishment and so accepted by the judge."

Judge Abraham L. Marovitz, who at the time had twenty-six years' experience as a judge, twelve of them on the federal bench, asked rhetorically in 1976: "Why then do I fear patent cases and things that go 'bump, in the night?' " He went on:

> When I drew a patent case, to be perfectly honest, I was less than overjoyed. I envisioned inordinate amounts of trial time, mountains of documents, reams of testimony, countless mysterious diagrams, and endless verbal duels between experts who speak in a foreign tongue and write in an alien language. I was overcome by a feeling that I was being compelled to perform in a role for which I was dreadfully ill prepared, and to witness the judicial system operating at its poorest. . . .
>
> What is it that distinguishes patent suits from antitrust, criminal, securities, and other federal matters? First and foremost is the time element. . . .
>
> I am not trying to be facetious when I draw this analogy: most cases are in English, but in patent litigation it is as if the judge is suddenly required to learn Chinese in order to decide a case. . . . Therefore, I do not believe that justice is best served by requiring patent clients and patent lawyers to try their cases before "patent-ignorant" judges.

Surprising as it may seem, there is only one federal district judge in the entire United States, William C. Conner of the Southern District of New York, who was a patent lawyer before becoming a judge. Even more astonishing, there were, in the entire history of the Court of Customs and Patent Appeals, only four judges who were formerly patent lawyers. Two of them, Chief Judge Howard T. Markey and Judge Giles

S. Rich, are now on the Court of Appeals for the Federal Circuit. Judge Markey is Chief Judge of the new Court. He has much knowledge and expertise not only in patents, but in numerous fields of law, having sat on many courts of appeals hearing a wide variety of cases.

A body of patent-experienced judges should be created at the federal district level by the appointment of at least one patent lawyer to the bench in every major metropolitan area. Patent cases would automatically be assigned either to presently sitting judges who want to hear patent cases, or to these patent-lawyers-turned-judges. Judge Marovitz has endorsed this suggestion, writing: "Such a system would free patent-inexperienced judges to uncrowd their overloaded dockets and would be a great benefit, not only to the just and speedy resolution of patent cases, but to the judicial system as a whole."

By creating a group of patent-experienced judges at the district level, it will eventually become possible to appoint the best from among them to the new Court of Appeals for the Federal Circuit. The goal should be for perhaps half the judges on that Court to have extensive patent backgrounds *prior* to appointment. This suggestion is made even though many nonpatent-lawyer judges on the new Court have and will become expert in patent cases, because of desire and experience. However, experience requires much time; there should be a reasonable mix of judges who were, and were not, previously patent lawyers.

A further benefit to be derived from the development of patent-experienced judges would be a shortening of trials and diminution of the abuse of the discovery (depositions, and so forth) process, which almost everyone in the legal field agrees has gotten out of hand. It is discovery—not trials—that makes patent litigation so expensive. In a discussion relating primarily to patent lawsuits, District Judge Edward Weinfeld called excessive discovery a "tremendous, extraordinary waste of time, and the imposition of very heavy expense, both upon your own immediate client, and the client of the opposing side."

Another judge declared:

> Patent attorneys overdo discovery more than any other segment of the Bar. . . . Once upon a time, . . . the whole philosophy of the new discovery rule was to eliminate the trial by combat aspect of the trial and let everybody go into trial knowing what the other side knew, so that there were no . . . surprises. This was supposed to serve justice better, but it led to an obvious area of abuse and that is the abuse of the solvent litigant discovering the insolvent litigant right out of the case and forcing him into settlement because he can't afford to go on. . . . The answer . . . is that the judges have been much too lax. They have not sufficiently controlled or supervised discovery, and they've permitted too much excessive discovery . . . and the process tends to be more endless in patent cases than any other kind.

Judges experienced in patent litigation and versed in technology will be in a far better position to know which matters are, and are not, of real importance in patent cases, and hence will be able to limit discovery, expedite trials, and reduce greatly the cost of lawsuits.

Knowledge and Hope

Until recently, many inventors, managers, and investors regarded patents as atomic cannons easily used to stop copyists. The thought that a judge might invalidate a patent never entered their heads. Typical thinking was: The government issued the patent, didn't it, so how could it be invalid? When I explained the real situation to such persons, looks of discouragement would appear on their faces. The discouragement could be countered, but only in part, by my explanation that very few patents are sued on to any extent; most infringers would rather pay or quit than hassle.

In short, through 1976, full knowledge of the patent system was not compatible with the full *hope* that inventors and investors need so badly. They have many severe problems; incorrect invalidations are the last straw. The following quote from a speech made in 1977 by a noted and successful independent inventor, Jacob Rabinow, emphasizes the importance of hope:

> . . . you should fix the odds at least so that the game is worth playing. That doesn't mean you have to make every inventor rich. . . . [But] it seems to me that people who have an influence on these things, the Department of Justice, the Congress, the patent bar, and so on, should realize that *unless they keep hope alive that sometimes we can break through and finally make it, we will not continue doing it*, and if the odds are too much against us, we will quit. And if we quit, the country goes down with us, because we cannot have a rich country like the United States unless we have something to sell and export. . . . (italics mine)

Since the time of the media awakening, beginning in 1980, many inventors have gone to the other extreme, regarding patents not as atomic cannons but rather as being relatively worthless. A small independent-inventor client of mine recently remarked, "My friends say that patents aren't worth a damn." How wrong they are! They would have been wrong even during the worst of times, and are in far greater error today. Such statements (and the old saw "a patent is just a ticket to a lawsuit") illustrate at least as much lack of information, and misinformation, as is illustrated by the viewing of patents as atomic cannons.

Thanks to the great stimulants introduced into the patent system by the recent Acts of Congress, the recent change in the attitude of the courts, and the turnaround in the Antitrust Division, full knowledge of the patent system by an inventor, investor, or entrepreneur is no longer incompatible with maintaining full hope. Patents are having a rebirth. They do and will give exclusivity with respect to reasonably good inventions, even when an infringer fights hard. And exclusivity for a limited period, as founded on the Constitutional plan, is one of the essentials for major progress in the useful arts.

ADDENDUM

On May 24, 1983, the Supreme Court of the United States issued a strongly pro-patent decision and opinion. The decision was *unanimous.* Speaking through Justice Marshall, the Court said, "prejudgment interest [in patent infringement actions] should ordinarily be awarded absent some justification for withholding such an award."

Interest is added to the damages given a victorious patent owner, and (under the decision) starts to build up as soon as infringement occurs. In the case at bar, the amount that the defendant (General Motors) had to pay the patent owner (Devex Corporation) was increased by millions of dollars because of interest charged for the period prior to judgment.

APPENDIX

The Eleven Federal Judicial Circuits

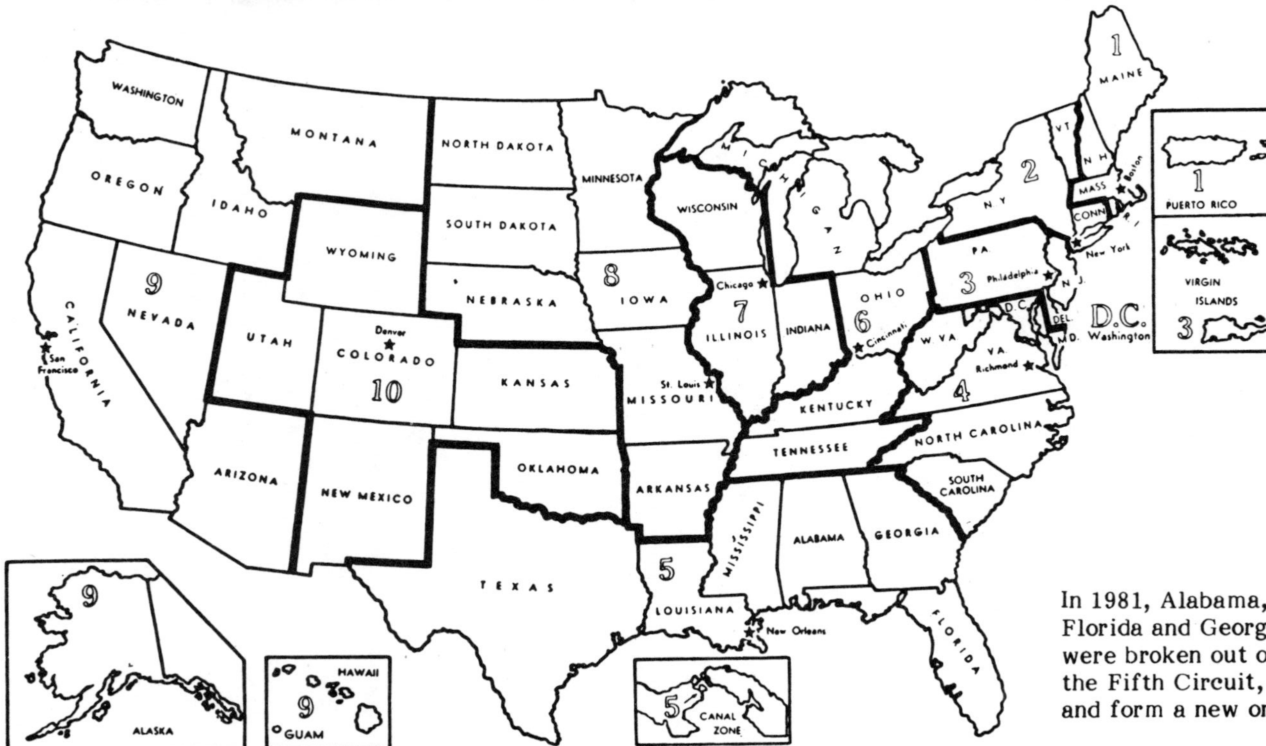

In 1981, Alabama, Florida and Georgia were broken out of the Fifth Circuit, and form a new one.

Used by permission of West Publishing Company.

SOURCE NOTES

Explanation

The numbers on the left below are page numbers of this book where each reference appears. The two words following each page number are the last two words of the sentence(s) to which the source note refers. When a note is not given, the source of either quoted material or stated information is a personal interview or public and common knowledge.

The first reference to a source gives its citation in full, unless the citation is given in the "Abbreviations" below. Subsequent references to the same source appear in shortened form.

Abbreviations

APLA Bulletin: *American Patent Law Association Bulletin,* Arlington, Va.
APLA Journal: *American Patent Law Association Quarterly Journal,* Arlington, Va.
F. and F. 2d: *Federal Reporter* and *Federal Reporter, Second Series,* respectively. Reports of cases argued and determined in the U.S. Courts of Appeals, U.S. Court of Claims, and U.S. Court of Customs and Patent Appeals (St. Paul, Minn.: West Publishing Co., 1982).
F. Supp.: *Federal Supplement.* Reports of cases argued and determined in the U.S. District Courts and certain other courts (St. Paul, Minn.: West Publishing Co., 1982).
JPOS: *Journal of the Patent Office Society,* Arlington, Va.
Nonobviousness: *Nonobviousness–The Ultimate Condition of Patentability,* John F. Witherspoon, ed. Reprinted by permission. Copyright © 1978, 1980 by the Bureau of National Affairs, Inc., Washington, D.C.
PTCJ: *The Bureau of National Affairs' Patent, Trademark & Copyright Journal.* Copyright © on the stated years of publication by the Bureau of National Affairs, Inc., Washington, D.C.
U.S.: *United States Reports.* Reports of cases adjudged in the U.S. Supreme Court (Washington, D.C.: U.S. Government Printing Office, 1982).
USPQ: *The United States Patents Quarterly.* Reports of cases relating to patents, trademarks, and copyrights (Washington, D.C.: The Bureau of National Affairs, Inc., 1982).

Preface

vii the mountain: Tom Arnold, "Licensing Climate in the U.S.A.," *Les Nouvelles,* Journal of the Licensing Executives Society, vol. 16 (September 1981), p. 188.
vii its functioning: Howard I. Forman, "America's Bicentennial and the American Patent System," *JPOS* 61 (August 1979), pp. 504, 505.

Chapter One: The Obvious Girdle

2 rags-to-riches tradition: *The Olga Company 1981 Annual Report,* 7900 Haskell Avenue, Van Nuys, California, p. 1.
3 built them: "The Mind's Eye: Nonverbal Thought in Technology," *Science* (26 August 1977), pp. 827–836.
6 was fashionable: Vanity Fair Mills Inc. v. Olga Company (Inc.), Appeal No. 74-1559, *Joint Appendix,* United States Court of Appeals for the Second Circuit, p. 335a.
8 volume zooming: *Women's Wear Daily* (24 March 1966), p. 26.
8 near nothing: *Joint Appendix,* p. 323a.

9 Tumm-ee Breef: Patents No. 3,142,300 and 3,142,301, issued July 28, 1964.
10 it down: *Joint Appendix,* p. 294a.
10 garment structure: Vanity Fair Mills, Inc. v. Olga Company (Inc.), 396 *F. Supp.* 1233, at 1240 (1974).
10 transitory fashion: Ibid., p. 1237.
10 their . . . discoveries: *Constitution of the United States,* Article I, Section 8, Cl. 8.
15 the business: *Women's Wear Daily* (26 October 1967), p. 15.
16 tummy control: Vanity Fair Mills, Inc. v. Olga Company (Inc.), Case No. 67 Civ. 4181, Deposition taken 19 November 1968, United States District Court, Southern District of New York, p. 15.
16 brief style: Ibid., another Deposition taken 19 November 1968, p. 16.
16 tricot gusset: Vanity Fair Mills Inc. v. Olga Company (Inc.), Appeal No. 74-1559, *Exhibit Volume,* U.S. Court of Appeals for the Second Circuit, pp. E103, E160.
16 largest-selling brief: 396 *F. Supp.* 1233, at 1238 (1974).
16 leg opening: *Joint Appendix,* p. 237a.
16 anything new: Case No. 67 Civ. 4181, another Deposition taken 19 November 1968, pp. 46–48.
16 asked Vanity Fair: *Exhibit Volume,* p. E164.
16 following points: Vanity Fair Mills Inc. v. Olga Company (Inc.), Case No. 74-1559, *Plaintiff-Appellant's Brief,* U.S. Court of Appeals for the Second Circuit, p. 34.
17 I refer: *Exhibit Volume,* pp. E161–E163.
17 to settle: *United States Code Annotated,* Title 28, Section 2201 (St. Paul, Minn.: West Publishing Company, 1982).
19 girdle business: *Joint Appendix,* p. 348a.
20 without them: Ibid., pp. 260a, 296a, 299a, 300a, 354a.
20 two garments: 396 *F. Supp.* 1233, at 1239 (1974).
20 six years: Ibid., pp. 1238, 1241, 1242.
21, 22 same results: Vanity Fair Mills, Inc. v. Olga Company (Inc.), 510 *F. 2d* 336, at 339, 340 (1975).

Chapter Two: The Incredible Tangle

23 William Newton: "Report from the Select Committee on the Law Relative to Patents for Inventions, 1829," quoted in footnote No. 9: Eric Robinson, "James Watt and the Law of Patents," *Technology and Culture* (April 1972), p. 117.
23, 24 conflicting doctrines: John Jewkes, David Sawers, and Richard Stillerman, *The Sources of Invention,* 2nd ed., paperback (New York: W. W. Norton & Company, Inc., 1969), pp. 72, 188.
24 is genius: Robert C. Post, "The American Genius," *The Smithsonian Book of Invention* (New York: W.W. Norton & Company, 1978), p. 23.
25 similar cases: P. J. Federico, "Origin and Early History of Patents," *JPOS* 11 (October 1929), pp. 294, 295.
25 much loss: Robert A. Choate and William H. Francis, *Cases and Materials on Patent Law* (St. Paul, Minn.: West Publishing Co., 1973), p. 60.
25 be questioned: Karl Fenning, "The Origin of the Patent and Copyright Clause of the Constitution," *JPOS* 11 (October 1929), p. 443.
26 at all: First U.S. Patent Grant, issued July 31, 1790. Also, see P. J. Federico, "Operation of the Patent Act of 1790," *JPOS* 18 (April 1936), pp. 237, 251.
26 improvement thereon: Anthony W. Deller, *Walker on Patents,* Deller's Edition, Volume 1 of 4 (New York: Baker, Voorhis and Company, 1937), pp. 531–535. Reprinted from *Walker on Patents* by A. W. Deller by special permission. Copyrighted by Baker, Voorhis and Co., Inc.
26 an invention: Hotchkiss et al. v. Greenwood et al., 52 *U.S.* 248 (1850).

27 the inventor: Ibid., p. 267.
27 never patentable: Ibid., p. 266.
27 all proportion: Giles S. Rich, "Escaping the Tyranny of Words—Is Evolution in Legal Thinking Impossible?" *JPOS* 60 (May 1978), p. 293.
28 prose resulting: Giles S. Rich, "The Principles of Patentability," *JPOS* 42 (February 1960), p. 88.
28 a patent: 52 *U.S.* 248, at 266 (1850).
29 a patent: Smith v. Goodyear Dental Vulcanite Company, 93 *U.S.* 486, 496 (1876).
29 set forth: Edison Electric Light Co. v. United States Electric Lighting Co., 52 *F.* 300, at 301, Circuit Court of Appeals, Second Circuit (1892).
29 earlier patent: Ibid., p. 304.
30 to come: Van Heusen Products, Inc. et al. v. Earl & Wilson. Same v. Cluett, Peabody & Co., Inc., 300 *F.* 922, at 929 (1924).
30 was subjective: Giles S. Rich, "Laying the Ghost of the 'Invention' Requirement," *APLA Journal* 1 (December 1972), p. 31.
30 than wood: Josephine H. Peirce, *Fire on the Hearth* (Springfield, Mass.: Pond-Ekberg Co., 1951), pp. 126–141.
31 burner stove: illustration taken from Hailes v. Van Wormer, 87 *U.S.* 353 (1873).
32 Glory stoves: D. G. Littlefield, *Theory of the Base-burning Stove and the Origin of the Morning Glory* (Albany, New York: Littlefield Stove Mfg. Co., 1871).
32 honest purpose: George R. Howell and J. Tenney, *Bicentennial History of Albany* (New York: W. W. Munsell & Co., 1886), p. 568.
32 per year: Edward Fitzgerald, *The Manufacturers of Albany* (Albany, New York: The Argus Co., 1872).
33 economical manner: Hailes and Treadwell v. Van Wormer et al., *Appendix A* of the record transmitted to the U.S. Supreme Court, (1870), p. 5.
33 same devices: 87 *U.S.* 353, at 368 (1873).
33 nonsense rhyme: Edmund W. Kitch, "Graham v. John Deere Co.: New Standards For Patents," *JPOS* 49 (April 1967), p. 265.
34 Faber pencils: Reckendorfer v. Faber, 92 *U.S.* 347 (1875).
34 and Eraser: United States Patent No. 19,783, dated March 30, 1858.
34 tack-drawer: 92 *U.S.* 347, at 356 (1875).
34 separate elements: Ibid., p. 357.
35 whole log: Ibid.
35 therefore, patentable: Ibid.
35 are forgotten: Rich, *APLA Journal* 1 (December 1972), p. 30.
35 question, cooperate: Pickering v. McCullough, 104 *U.S.* 310, at 317, 318 (1881).
36 checkout counter: Great Atlantic & Pacific Tea Co. v. Supermarket Equipment Corp., 340 *U.S.* 147 (1950).
36 the patent: Bradley v. Great Atlantic & Pacific Tea Co., 78 *F. Supp.* 388 (1948).
36 of appeals: Great Atlantic & Pacific Tea Co. v. Supermarket Equipment Corporation, 179 *F. 2d* 636 (1950).
37 checkout counter: The Great Atlantic & Pacific Tea Company v. Supermarket Equipment Corporation, *Brief for Respondent* on Writ of Certiorari, Plaintiff Exhibit No. 25, in the Supreme Court, Case No. 32 (1950), p. 28.
37 elements together: 340 *U.S.* 147, at 150 (1950).
38 but confusion: Ibid., pp. 150, 151.
38 severe test: Ibid., p. 152.
39 pool table: 179 *F. 2d* 636, at 639 (1950).
40 ever done: 340 *U.S.* 147, at 152 (1950).
42 stimulate invention: Atlantic Works v. Brady, 107 *U.S.* 192, at 200 (1882).
42 all favor: Ibid.

Chapter Three: The Fallout from A&P

43 Learned Hand: Reiner v. I. Leon Co., Inc., 285 *F. 2d* 501, at 503 (1960).
43 previous day: Rich, *APLA Journal* 1 (December 1972), p. 32. Also, see *The New York Times* (5 December 1950), p. 33.
43 Patent Office: P. J. Federico, "Origins of Section 103," *Nonobviousness* (see Note Abbreviations for full citation), pp. 1:103–1:105. See Giles S. Rich, "P.J. (Pat) Federico and His Works," *JPOS* 64 (January 1982), pp. 3–11.
44 checkout counter: Rich, *APLA Journal* 1 (December 1972), p. 32.
44 is not: Ibid., p. 33.
44 is reversed: 340 *U.S.* 147, at 154 (1950).
44 old elements: Ibid., p. 152.
44 held valid: Parks v. Booth, 102 *U.S.* 96, at 104 (1880).
44 new assemblage: 285 *F. 2d* 501, at 503 (1960).
45 an art: Rubber Company v. Goodyear, 76 *U.S.* 788, at 795 (1869); and Cuno Engineering Corp. v. Automatic Devices Corp., 314 *U.S.* 84, at 92 (1941).
45 public domain: 314 *U.S.* 84, at 91 (1941).
45 the world: Marconi Wireless Telegraph Company of America v. United States, 320 *U.S.* 1, at 62, 63 (1942).
46 big results: Ibid., p. 74.
46 revised form: P. J. Federico, "Commentary on the New Patent Law," *United States Code Annotated*, Title 35, Patents, pp. 22, 23. Also, see Rich, *APLA Journal* 1 (December 1972), pp. 33, 37.
46 matter pertains: *United States Code Annotated*, Title 35, Section 103.
46 or catchwords: Federico, *United States Code Annotated*, Title 35, p. 20.
46 in granite: William O. Douglas, *The Court Years 1939–1975* (New York: Random House, 1980), p. 8.
47 our predilections: Ibid.
47 highly competitive: Ibid.
47 relevant evidence: Edmund A. Godula, "Judge Learned Hand and the Concept of Invention," *Idea* 9 (Spring 1965), pp. 159–181. *Idea*, the Journal of Law and Technology, is published by the PTC Research Foundation of The Franklin Pierce Law Center (Concord, NH). Judge Hand's cases are well summarized in this article.
47 as possible: B.G. Corporation v. Walter Kidde & Co., Inc., 79 *F. 2d* 20, at 22 (1935).
47 with it: 285 *F. 2d* 501, at 504 (1960).
47 the Court: Kathryn Griffith, *Judge Learned Hand and the Role of the Federal Judiciary* (Norman: University of Oklahoma Press, 1973), p. 9.
48 used it: 285 *F. 2d* 501, at 504 (1960).
48 was made: *The New York Times* (30 August 1953), in "Letters to the Times." Copyrighted 1953 by The New York Times Company.
49 be sustained: 510 *F. 2d* 336, at 340 (1975).
50 make patentability: 340 *U.S.* 147, at 153 (1950).
50 hands on: Jungersen v. Ostby & Barton Co. et al., 335 *U.S.* 560, at 571, 572 (1948).
50 our predilections: Douglas, *The Court Years*, p. 8.
51 of patentability: Goodyear v. Ray-O-Vac, 321 *U.S.* 275, at 279 (1944).

Chapter Four: One Up, Two Down in the Supreme Court

53 Thomas Jefferson: A letter to Dr. Benjamin Waterhouse, Monticello, March 3, 1818. *The Life and Selected Writings of Thomas Jefferson*, A. Koch and W. Peden, eds. (New York: The Modern Library, Random House, Inc., 1972), p. 685.

54 to patents: 285 *F. 2d* 501, at 503 (1960). Also, see a famous case, Lyon v. Bausch and Lomb Optical Co., 224 *F. 2d* 530 (1955), opinion by Judge Learned Hand.
54 Government hornswoggling: "The Law," *Time* (4 March 1966), p. 61.
55 and painting: United States v. Bert N. Adams et al., *Transcript of Record* (of the proceedings in the U.S. Court of Claims), in the Supreme Court of the United States, Case No. 55 (October Term, 1965), p. 39.
55 Volta believed: Transcript of Oral Argument in United States v. Adams. Appendix of *Nonobviousness*, p. A:8.
55 be corrosive: *Transcript of Record,* Case No. 55, p. 40.
56 feeble current: Ibid., p. 42.
57 at first: Ibid., pp. 44–47.
57 check them: Ibid., p. 107.
57 Inventors Council: Ibid., p. 411.
57 Signal Corps: United States of America v. Bert N. Adams et al., in the Supreme Court of the United States (October Term, 1964), from *Petition for a Writ of Certiorari* to the United States Court of Claims, finding of fact no. 40, pp. 35, 36.
58 due time: *Transcript of Record,* Case No. 55, pp. 77, 82.
58 improve it: *Petition for a Writ of Certiorari,* p. 5.
58 be needed: *Transcript of Record,* Case No. 55, pp. 402, 403.
59 partly overcome: Ibid., p. 219.
59 military fliers: *Time* (4 March 1966), p. 61.
59 repeated requests: United States v. Adams et al., 383 *U.S.* 39, at 44 (1966).
59 his invention: *Transcript of Record,* Case No. 55, pp. 67, 71.
59 of his invention: 383 *U.S.* 39, at 44 (1966).
60 economically impractical: E. M. Morse, "Water-Activated Batteries," presented at the Tenth Annual Battery Research and Development Conference (23 May 1956), *Transcript of Record,* Case No. 55, pp. 439, 440.
61 that effect: *Petition for a Writ of Certiorari,* p. 37.
61 Court's opinion: Charles D. Reed, "Some Reflections on Graham v. John Deere Co.," *Nonobviousness,* pp. 2:301–2:303.
62 was ridiculous: "Transcript of Oral Argument," *Nonobviousness,* pp. A:8,9.
62 the Justices: "John Reilly Speaks at Luncheon," *APLA Bulletin* (October-November 1966), p. 491.
63 of confidence: Reed, *Nonobviousness,* pp. 2:303, 2:304.
63 same system: 383 *U.S.* 39, at 41, 52 (1966).
63 the Bureaucrats: *Time* (4 March 1966), p. 61.
63 it did: *The Christian Science Monitor* (26 February 1966).
63 be done: *United States Congressional Record–Senate* (18 October 1967), S14909.
64 Adams batteries: *Respondents' Brief,* Case No. 55, p. 49.
64 under lock: Reed, *Nonobviousness,* p. 2:304.
64 it does: "Transcript of Oral Argument," *Nonobviousness,* pp. A:8,13.
65 prior art: 383 *U.S.* 39, at 50 (1966).
66 to emerge: Robert H. Rines, "Patents and the Stimulation of Innovation," *IEEE Transactions of Engineering Management* Em-15 (2 June 1968), p. 72.
66 neglected precedent: John A. Reilly, "A Post-AG PRO Look at United States v. Adams," *Nonobviousness,* p. 2:402.
66 for both: Graham et al. v. John Deere Co. of Kansas City et al. (together with Calmar, Inc. v. Cook Chemical Co.), 383 *U.S.* 1 (1966).
67 its meaning: Ibid., pp. 11, 12.
67 have relevancy: Ibid., pp. 17, 18.
67 patentable invention: Ibid., p. 17.
67 be applied: Ibid., p. 19.

67 not delivered: Ibid., pp. 12–17.
68 of litigation: Ibid., p. 21.
68 been obvious: Ibid., pp. 25, 26.
69 the units: Ibid., pp. 27, 28. Also, see 336 *F. 2d* 110, at 113 (1964).
69 a gadget: Calmar Inc., v. Cook Chemical Company, *Petition for a Writ of Certiorari,* Case No. 37, Supreme Court (October Term 1965), pp. 10, 18.
69 been obvious (first occurrence): *United States Code Annotated,* Title 35, Patents, Section 103.
69 commercial success: 336 *F. 2d* 110, at 114 (1966).
69 the art: 383 *U.S.* 1, at 36 (1966).
70 is determined: Ibid., p. 17.

Chapter Five: Damage in the Lower Courts

74 present controversy: L & A Products v. Britt Tech, 365 *F. 2d* 83, at 86 (1966).
74 was sustained: Myron Cohen, "Nonobviousness and the Circuit Courts of Appeals—Twenty-five Years in Review," *Nonobviousness*, p. 3:42.
74 commercial success: 510 *F. 2d* 336, at 339 (1975).
75 the year: John Brooks, *The Go-Go Years* (New York: Weybright and Tulley, 1973), pp. 27–29.
77 dead-silent background: "Revolution in Radio," *Fortune* (October 1939), p. 88.
78 move over: M. Randolph Preuss v. General Electric Co., Inc., *Appellant's Appendix*, Docket No. 31507, U.S. Court of Appeals for the Second Circuit, p. 108a.
78 can achieve: Harold Fletcher, "An Acoustic Illusion Telephonically Achieved," *Bell Laboratories Record* (May 1933), pp. 286–289.
78 very severe: "Evolution of a HiFi Broadcast," *FM Guide* (October 1966), p. 28.
79 New Jersey: Elliott M. Sanger, *Rebel in Radio: The Story of WQXR* (New York: Hastings House, 1973), p. 50.
81 binaural programs: *Electrical Engineering* (August 1953), p. 713.
81 this field: M. Rudolph Preuss v. General Electric Co., Inc., 269 *F. Supp.* 993, at 994, 995 (1967).
82 SCA operation: "Permission of FM Broadcast Stations to Transmit Stereophonic Programs on a Multiplex Basis," *Federal Register* 26 (25 April 1961), p. 3529.
83 as possible: *The Wall Street Journal* (19 April 1961), p. 5.
83 September 1966: 269 *F. Supp.* 993 (1967).
83 inventors lived: United States Patent 2,698,379.
83 FM broadcasting: M. Rudolph Preuss v. General Electric Co., Inc., *Appellant's Brief,* Docket No. 31507, U.S. Court of Appeals for the Second Circuit, p. 26.
84 on it: *Appellant's Appendix,* Docket No. 31507, p. 146a.
84 an afterthought: 383 *U.S.* 1, at 25 (1966). Also, see Robert A. Choate, "Invention and Obviousness—'Afterthoughts'—Reliance on the Features and Advantages Undisclosed at Original Filing," *JPOS* 49 (September 1967), pp. 619–636.
85 two decades: 269 *F. Supp.* 993, at 995 (1967).
85 his system: Ibid., p. 996.
85 was valid: Ibid., p. 998.
85 percent thereafter: *Appellant's Appendix,* Docket No. 31507, p. 101a.
85 upon here: *Appellant's Brief,* p. 26.
86 A&P ruling: Ibid., p. 55.
86 of each: Ibid., p. 43.
86 subject matter: Ibid., p. 53.
86 giant stride: Ibid., p. 50.
86 of hoopla: Ibid., p. 60.

86 and band-width: Preuss v. General Electric, *Brief for Plaintiff-Appellee,* Docket No. 31507, U.S. Court of Appeals for the Second Circuit, pp. 9–13.
86 Crosby's shift: Ibid., pp. 59, 60.
86 the solution: Ibid., p. 36.
86 had invented: Preuss v. General Electric, 392 *F. 2d* 29, at 31 (1968).
86 an afterthought: Ibid., p. 36.
87 a burden: Ibid., p. 33.
87 financial resources: Preuss v. General Electric, *Petition for Writ of Certiorari,* in the Supreme Court (October Term 1968), p. 8.
87 energy gone: Lawrence Lessing, *Man of High Fidelity: Edwin Howard Armstrong* (Philadelphia: J.B. Lippincott Co., 1969), pp. 246–248.

Chapter Six: A Big Winner

91 him happy: *PIC Magazine* (November 1946), p. 23.
94 corresponding method: Patents 2,685,536 and 2,794,417.
95 by hand: "Your Life Tomorrow," *Collier's Magazine* (7 July 1945).
95 without clogging: Patents 2,893,893, 2,893,894, 2,247,963, and 2,334,648.
95 valuable patents: "Ransburg Electro-Coating Corporation," *Indiana Business and Industry* (March 1964), p. 6.
96 the years: "Telling the Judge Firm's Old Story," *The Indianapolis Business-World News* (20 June 1968).
97 of cases: Robert V. Bruce, *Alexander Graham Bell and the Conquest of Solitude* (New York: Little, Brown & Co., 1973), p. 271.
98 royalty payments: Binks Manufacturing Company v. Ransburg Electro-Coating Corp., 122 *USPQ* 74, at 76, 77 (1959).
99 was invalid: Binks Manufacturing Company v. Ransburg Electro-Coating Corp., *Brief and Supplemental Appendix for Plaintiff-Appellant,* Case No. 12737, in the United States Court of Appeals for the Seventh Circuit (1960), p. 12. Also, see United States Patent 1,958,406.
99 was joined: 122 *USPQ* 74, at 76, 77 (1959).
99 it rest: Binks Manufacturing Company v. Ransburg Electro-Coating Corp. Deposition of William Darrah, in Plaintiff's Appendix, *Plaintiff's Exhibit 0000,* Case No. 12737, in the U.S. Court of Appeals for the Seventh Circuit (1960), p. 662.
99 record here: 122 *USPQ* 74, at 83 (1959).
99 its patents: Ibid., p. 91.
99 paint particles: Ibid., pp. 88, 89.
100 measuring invention: Binks Manufacturing Company v. Ransburg Electro-Coating Corp., *Brief for Defendant Appellee and Annotated Findings of Fact and Conclusions of Law,* in the United States Court of Appeals for the Seventh Circuit, p. 56.
100 this statement: Binks Manufacturing Company v. Ransburg Electro-Coating Corporation, 281 *F. 2d* 252, at 255, 256 (1960).
101 Research Institute: Ransburg Electro-Coating Corp. v. Nordson Corp., 293 *F. Supp.* 448, at 452 (1968).
101 Process patents: Ransburg Electro-Coating Corp. v. Ford Motor Company, 245 *F. Supp.* 308 (1965).
101 $2 million: Ransburg Electro-Coating Corp. v. Standard Container Company, 167 *USPQ* 426 (1970).
102 did it: "I Had to Find a Better Way," *The American Machinist* (25 April 1966).
103 called Cybotech: *Ransburg: The Decade of Industrial Opportunity,* sponsored by Indianapolis Securities Traders Association, a booklet (4 November 1981).

Chapter Seven: The Road Toward Confusion

105 George Washington: Deller, *Walker on Patents,* Deller's Edition vol. 1 (1937), p. 32.
107 old material: Pavement Salvage Co., Inc. v. Anderson's Black Rock, Inc., 404 *F. 2d* 450, at 451 (1968).
107 for himself: Ibid., p. 453.
110 highly successful: Earl Whithycombe, "An Experiment," *California Highways and Public Works* (November–December 1950), pp. 24-27.
110 Reclaimix Inc.: *Sacramento Union* (8 March 1951).
110 behind it: *Los Angeles Daily News* (8 March 1951).
111 Neville's product: Pavement Salvage Company, Inc. v. Anderson's Black Rock, Inc., Civil Action 2927, United States District Court Southern District of West Virginia at Charleston, (19, 20 April 1965), p. 75.
113 the Hudson: S. C. Gilfillan, *Inventing the Ship: A Study of Inventions Made in Her History Between Floating Log and Rotor Ship* (New York: Follett, 1935).
113 after him: S. Colum Gilfillan, *Supplement to the Sociology of Invention* (San Francisco: San Francisco Press, Inc., 1971), p. 49.
114 antecedent process: 320 *U.S.* 1, at 62 (1943).
114 Gunther Schwank: United States Patent 2,775,294 (issued 25 December 1956).
114 two machines: United States Patent 3,055,280 (issued 25 September 1962).
114 foreign countries: Pavement Salvage v. Anderson's Black Rock, C. Action 2927, p. 78.
115 declared invalid: Anderson's Black Rock, Inc. v. Pavement Salvage Co., Inc., *Appendix IV,* Opinion of the District Court, S.D.W. Va., March 23, 1964, contained in Petition for Writ of Certiorari, in the Supreme Court of the United States, Case No. 45 (October Term, 1968), pp. A15–A26.
115 patent valid: 404 *F. 2d* 450 (1968).
115 heat treatment: Ibid., p. 453.
116 the experts: Ibid., p. 454.
116 not known: Ibid.
117 the patent: Anderson's Black Rock v. Pavement Salvage Co., 396 *U.S.* 57 (1969).
117 combination patents: Ibid., p. 60.
117 the paver: Ibid.
118 the art: Ibid.
118 Supermarket Corp.: Ibid., p. 61.
118 already patented: Ibid., pp. 61, 62.

Chapter Eight: The Cow Barn Disaster

121 Apollo XII: Edmond Rostand, *Cyrano de Bergerac,* translation by Lowell Bair (New York: New American Library, a Signet Classic, paperback 1972), p. 135.
122 Next case: Transcript, Walter Cronkite on "CBS Evening News" (20 April 1976).
124 ancient times: U.S. Patent No. 3,223,070, issued 14 December 1965.
124 and pitchforks: Bernard A. Sakraida v. AG PRO, Inc., Petition for Writ of Certiorari to the United States Court of Appeals for the Fifth Circuit, *Appendix,* in the Supreme Court, Case No. 75-110 (October Term, 1975), p. 107.
125 usual course: C. Witt, *Classic Mythology* (1883), pp. 119–120, quoted as footnote No. 1 in Sakraida v. AG PRO, 425 *U.S.* 273, at 275 (1976).
125 feeding stations: U.S. Patent No. 1,981,417 (issued 20 November 1934) and No. 1,981,418 (issued 20 November 1934).
126 cows' tails: U.S. Patent No. 1,968,564 (issued 31 July 1934).
126 off manure: U.S. Patent No. 2,081,947 (issued 1 June 1937).
127 of manure: U.S. Patent No. 2,233,766 (issued 4 March 1941).

128 the gutter: U.S. Patent No. 2,294,781 (issued 1 September 1942).
129 a manure tank: French Republic Patent No. 1,246,418 (issued 10 October 1960).
133 openable dam: U.S. Patent No. 3,223,070 (issued 14 December 1965).
134 few seconds: *Appendix,* Case No. 75-110, p. 61.
134 been used: Ibid., p. 108.
134 the past: Ibid., p. 109.
134 was going: *San Francisco Examiner* (28 June 1964).
135 pampered beauties: *The Weekly Calistogan* (4 June 1964).
135 a queen: *The Santa Rosa Press Democrat* (31 May 1964).
135 by AGPRO: AGPRO, Inc. v. Bernard A. Sakraida, Deposition taken 12 May 1969, Case No. Civil 68-118-EP, in the United States District Court for the Western District of Texas, El Paso Division, pp. 4, 19, 21, 22.
135 Bakersfield, California: Ibid., pp. 22–35, 45, 46.
136 of law: *U.S. Code Annotated,* Federal Rules of Civil Procedure, Title 23, Rule 56(c), p. 413.
136 Section 103: AGPRO, Inc. v. Bernard A. Sakraida, *Brief in Support of Defendant's Motion for Summary Judgment,* Civil Action No. 68-118-EP, in the U.S. District Court for the Western District of Texas, El Paso Division (5 November 1969).
136 being patentable: AGPRO, Inc. v. Bernard A. Sakraida, *Defendant's Supplemental Brief in Support of His Motion for Summary Judgment,* Civil Action No. 68-118-EP, in the United States District Court for the Western District of Texas, El Paso Division, (23 December 1969), p. 5.
137 December 8, 1969: *Appendix,* Case No. 75-110, p. 45.
137 summarily truncated: AG PRO, Inc. v. Bernard A. Sakraida, 437 *F. 2d* 99, at 101 (1971).
137 about it: AG PRO, Inc. v. Bernard A. Sakraida, *Appendix,* Case No. 72-1108, in the United States Court of Appeals for the Fifth Circuit, pp. 184, 185.
137 and open it: *Appendix,* Case No. 75-110, p. 79.
138 it evolved: Ibid.
138 they approached: Ibid., p. 82.
138 the past: Ibid., p. 101.
139 the offal: Ibid., p. 90.
139 to judgment: Ibid., p. 55.
140 this patent: AG PRO, Inc. v. Bernard A. Sakraida, 474 *F. 2d* 167, at 169 (1973).
140 to him: Ibid., p. 170.
140 faculties did: Ibid., p. 171.
140 dramatic result: Ibid., p. 173.
141 Supreme Court: 425 *U.S.* 273 (1976).
141 and others: Bernard A. Sakraida v. AG PRO, Inc., *Brief for Petitioner* on Writ of Certiorari, in the Supreme Court of the United States, Case No. 75-110 (October Term, 1975), pp. 2, 4, 28–30, 37, 42, 44, 45.
141 Black Rock: Bernard A. Sakraida v. AG PRO, Inc., *Brief for Respondent* on Writ of Certiorari, in the Supreme Court of the United States, Case No. 75-110 (October Term, 1975), pp. 18, 22–25, 38.
141 patentable invention: Ibid., pp. 9, 33.
141 and invention: Eibel Co. v. Paper Co., 261 *U.S.* 45, at 68 (1923).
141 patent invalid: *Appendix,* Case No. 75-110, pp. 174–267.
141 due diligence: AG PRO, Inc. v. Sakraida, 512 *F. 2d* 141 (1975).
142 non-toilet-trained cows: Bernard A. Sakraida v. AG PRO, Inc., *Transcript* (by Hoover Reporting Co., Inc., Washington, D.C.), in the Supreme Court of the United States, Case No. 75-110 (3 March 1976), p. 26.

142 Deere, A&P: Ibid., p. 4.
142 concrete floor: Ibid., pp. 5, 6.
142 flush system: Ibid., p. 12.
143 the floor: Ibid., pp. 12, 13. Hercules was referred to previously by Kolisch in the 1971 trial. See *Appendix,* Case No. 75-110, p. 57.
143 certainly does: *Transcript,* p. 13.
143 sanitary requirements: Ibid., p. 29.
143 wave-generating mechanism: Ibid., pp. 33, 34.
144 will work: Ibid., pp. 37, 38.
144 patent validity: 425 *U.S.* 273, at 274 (1976).
144 ancient times: Ibid. p. 275.
144 old elements: Ibid., p. 281.
145 Pavement Co.: Ibid., p. 282.
145 Supermarket Corp.: Ibid., pp. 282, 283.
146 of speech: Bernard A. Sakraida v. AG PRO, Inc., *Brief on Behalf of the American Patent Law Association,* as Amicus Curiae, in Support of Respondent's Petition for Rehearing, in the Supreme Court, Case No. 75-110 (October Term, 1975), p. 15.
146 A&P concepts: Dann, Commissioner of Patents and Trademarks v. Johnston, 425 *U.S.* 219 (1976).
147 of cooperation: Lincoln Engineering Co. v. Stewart-Warner Corp., 303 *U.S.* 545 (1938). Also, see Pickering v. McCullough, 104 *U.S.* 310 (1881).
148 distinct trauma: *PTCJ* (4 March 1976), pp. A2–A4, A13, and Special Supplement. Also, see *PTCJ* (3 October 1974), pp. A2, A3.

Chapter Nine: Why the Unequal Balance?

149 Benjamin N. Cardozo: Margaret E. Hall, Ed., "The Nature of the Judicial Process," *Selected Writings of Benjamin Nathan Cardozo* (New York: Fallon Publications, 1947), p. 139.
150 next Parliament: Deller, *Walker on Patents,* Deller's Edition Vol. 1, "An Historical Review of the Origin of Monopolies" (1937), p. 7.
150 or execution: Ibid., pp. 18, 19.
150 such manufactures: Ibid., pp. 19, 20.
150 are monopolies: Sears, Roebuck v. Stiffel, 376 *U.S.* 225, at 229, 230 (1964).
151 good side: Giles S. Rich, "The Vague Concept of 'Invention' As Replaced by Sec. 103 of the 1952 Patent Act," *JPOS* 46 (December 1964), pp. 858, 859.
152 should not: Graham et al. v. John Deere Co. of Kansas City et al., 383 *U.S.* 1, at 9, 10 (1966).
152 exclusive right: Paul L. Ford, Ed., "Draft of a Bill to Promote the Progress of the Useful Arts," *The Writings of Thomas Jefferson,* Vol. V (New York: G.P. Putnam's Sons, 1895), p. 279.
152 of patentability: 383 *U.S.* 1, at 9 (1966).
152 fourteen years: H.A. Washington, "A Letter to Mr. Pictet," *The Writings of Thomas Jefferson,* Vol. IV (New York: John C. Riker, 1857), pp. 462, 463.
152 December 1836: Bowen C. Dees, "An Institution's Love Affair with Inventions and Patents," *APLA Bulletin* (June 1976), p. 352.
152 of Canvass: Edmund Burke, Commissioner of Patents, Compiler, *List of Patents for Inventions and Designs Issued by the United States from 1790 to 1847,* with the patent laws and notes of decisions of the courts of the United States for the same period (Washington, D.C.: J. & G. S. Gideon, printer, 1847).
152 patents granted: Dumas Malone, *Jefferson and His Time,* Vol. Two, *Jefferson and the Rights of Man* (Boston: Little, Brown and Company, 1951), p. 283. Also, see Federico, *JPOS* 18 (April 1936), p. 246.

153 to ours: H. A. Washington, "A Letter to Mr. Isaac McPherson," p. 182.
154 and generously: *The Autobiography of Benjamin Franklin* (New York: Washington Square Press, Pocket Books, paperback, 1955), p. 143.
154 possible pirates: "The Protection by Patents of Scientific Discoveries," A Report of the Committee on Patents, Copyrights and Trade Marks by Joseph Rossman, Chairman; published in Supplement to *Science* 79 (January 1934), pp. 9, 10.
155 in 1860: Joseph Rossman, "Abraham Lincoln the Inventor," *JPOS* 10 (May 1928), pp. 292, 314–320.
155 is clear: Theodore L. Bowes, "The Patent-Antitrust Law Interface: How Should It Be Defined?" *Idea* 18 (Summer 1976), pp. 27, 28.
155 patent laws: Stephen P. Strickland, Ed., *Hugo Black and the Supreme Court* (Indianapolis: The Bobbs-Merrill Company, Inc., 1967), pp. 64, 65. Also, see Vern Countryman, *The Judicial Record of Justice William O. Douglas* (Cambridge, Mass.: Harvard University Press, 1974), pp. 328–341.
155 Patents Company: Robert Conot, *A Streak of Luck* (New York: Seaview Books, 1979), pp. 395, 396.
155 were illegal: Motion Picture Patents Company v. Universal Film Manufacturing Company et al., 243 *U.S.* 52 (1917).
156 manufactured product: Hartford-Empire Co. et al. v. United States, 323 *U.S.* 386, at 400 (1945).
156 in 1965: Walker Process Equipment, Inc. v. Food Machinery & Chemical Corp., 382 *U.S.* 172 (1965).
157 a patent: Ibid., p. 174.
157 the patentee: Lawrence B. Dodds, "Fraud on the Patent Office," *JPOS* 56 (June 1974), p. 346.
158 patent bar: Frederick B. Lacey, " A Federal District Judge's Views on Patent Reisue, Protest and Duty of Disclosure," *JPOS* 60 (September 1968), p. 350.
158 else could: *Night Key*, a Universal Picture Production distributed by Screen Gems.
159 a myth: Gilfillan, *Supplement to the Sociology of Invention*, pp. 29–31. Also, see "Materials Relating to Continuing Studies of Technology, Economic Growth, and the Variability of Private Investment Presented for Consideration of the Joint Economic Committee, Congress of the United States," *Invention and the Patent System*, 88th Congress, 2d Session, (Washington, D.C.: U.S. Government Printing Office, December 1964), pp. 98–102.
159 all about: Vannevar Bush, *Pieces of the Action* (New York: William Morrow and Company, Inc., 1970), p. 201.
159 dangerous land: Abe Fortas, "The Patent System in Distress," *JPOS* 53 (December 1971), p. 810. This article was originally printed in *Idea* 14 (Winter 1970–71).
160 Super Ultrex: Robert Levy, "Innovate or Replicate?" *Dun's Review* (June 1980), p. 87.
160 market competition: Rich, *JPOS* 46 (December 1964), p. 858.
161 will build: George W. Whitney, "President's Page," *APLA Bulletin* (September–October 1981), p. 491.
161 or development: Richard F. Carr, "Our Patent System Works," *JPOS* 42 (May 1960), pp. 317, 318.
162 and Company: John H. Dessauer, *My Years with Xerox* (New York: Doubleday & Company, Inc., 1971), pp. 23–27.
162 his invention: SCM Corp. v. Xerox Corp., 463 *F. Supp.* 983, at 992 (1978).
162 he said: "Chester Carlson Receives Inventor of the Year Award," *JPOS* 47 (May 1965), p. 287.
162 General Electric: Dessauer, p. 21.
163 Patent System: *JPOS* 47 (May 1965), p. 288.

163 when patented: 645 *F. 2d* 1195, at 1198 (1981).
163 thirty-six companies: Ibid.
163 corporate success: 463 *F. Supp.* 983, at 992 (1978).
163 working capital: Transcript of *Wall Street Week* with Louis Rukeyser, "Premature Growth Stocks," No. 751 (16 June 1978), p. 17.
164 Carlson did: Richard N. Currant, *The Typewriter and the Men Who Made It* (Champaign, Illinois: The University of Illinois, 1954), pp. 17–19.
164 Latham Sholes: Ibid., pp. 16, 17.
164 was made: Ibid., pp. 17, 29, 33, 38, 48, 53, 58, 59.
164 is famous: Ibid., pp. 65, 66.
164 the selling: Ibid., p. 65.
164 royalty-paying basis: Ibid., p. 82.
164 foreign-made machines: *Los Angeles Times* (25 December 1981), Part IV, p. 2.
168 that development: "Investigation of Concentration of Economic Power, *Final Report and Recommendations of the Temporary National Economic Committee,* 77th Congress, 1st Session, Senate, Document No. 35 (Washington, D.C.: United States Government Printing Office, 1941), p. 362.
169 Haloid Co.: "Concentration, Invention, and Innovation," Part 3, *Hearings Before the Subcommittee on Antitrust and Monopoly* of the Committee on the Judiciary, United States Senate, 89th Congress, 1st Session (1965), pp. 1115, 1116.
169 been overwhelmed: David F. Noble, *America by Design, Science, Technology, and the Rise of Corporate Capitalism* (New York: Alfred A. Knopf, 1977), e.g., p. 97.
169 inflation rates: Tom Alexander, "The Right Remedy for R&D Lag," *Fortune* (25 January 1982), p. 61.
170 become significant: Edwin H. Land, "The Role of Patents in the Growth of New Companies," *JPOS* 41 (July 1959), p. 502.
170 industrial climate: Vannevar Bush, "Some Proposals for Improving the Patent System," *JPOS* 38 (January 1957), p. 43.
171 universally recognized: Conot, pp. 99–104.
171 single sound: Ralph Stein, *The Great Inventions* (Chicago: The Ridge Press and Playboy Press, 1976), pp. 244, 245.
171 typewriter inventions: Michael H. Adler, *The Writing Machine* (London: George Allen & Unwin Ltd., 1973), p. 367.
171 numerous countries: Ibid., pp. 150–157.
172 step forward: Harry X. Bergman v. Aluminum Lock Shingle Corporation of America, 251 *F. 2d* 801, at 809 (1958).
172 readjustments required: Harries et al. v. Air King Products Co., Inc., 183 *F. 2d* 158, at 162 (1950).
173 then given: Laurence B. Dodds and Francis W. Crotty, "The New Doctrinal Trend," *JPOS* 30 (February 1948), p. 83.
174 an invention: Calmar, Inc. v. Cook Chemical Company, *Petition for a Writ of Certiorari,* in the Supreme Court, Case No. 37 (October Term 1965), p. 18.
174 matter pertains: *United States Code Annotated,* Title 35, Section 103.
174 the art: 383 *U.S.* 1, at 36 (1966).
175 not true: John A. Reilly, "A Post AGPRO Look at United States v. Adams," *Nonobviousness,* p. 2:406.
177 sole dissenter: Frank E. Johnston, "Mr. Justice Black on Patents," *JPOS* 54 (September 1972), p. 563.
177 liberty ships: Graver Tank et al. v. Linde Air Products 339 *U.S.* 594 (1950).
178 enterprise system: Ibid., p. 617.
178 concurring opinion: 340 *U.S.* 147, at 152–158 (1950).
178 voted against: Brenner v. Manson, 383 U.S. 519 (1966).

178 pivotally important: John A. Reilly, *APLA Bulletin* (October–November 1966), pp. 489–490.
180 scientific knowledge: 340 *U.S.* 147, at 154 (1950).
180 to patents: 383 *U.S.* 1, at 6 (1966).
181 useful arts: Rich, *JPOS* 60 (May 1978), pp. 282, 283.
182 that point: Reckendorfer v. Faber, 92 *U.S.* 347, at 351 (1875).
182 back inventors: Snow v. Commissioner of Internal Revenue, 416 *U.S.* 500 (1974).

Chapter Ten: Vastly Better Patent Pending

186 *Post* headline: Peter Elkind, "Patent Office Fails As Mother of Invention," *The Washington Post* (18 August 1980). Reprinted with permission of *The Washington Post*, copyright 1980.
186 it moribund: Ibid.
186 December 5, 1980: Transcript of television program, "Patently Absurd," *NBC Magazine* (5 December 1980).
188 invalidate patents: Ibid.
189 another firm: Sears, Roebuck & Co. v. Stiffel Company, 376 *U.S.* 225 (1964).
189 lighting fixture: Compco Corp. v. Day-Brite Lighting, Inc. 376 *U.S.* 234 (1964).
190 patent laws: 376 *U.S.* 225, at 231 (1964).
190 its benefits: Lear, Inc. v. Adkins, 395 *U.S.* 653 (1969).
190 be muzzled: Ibid., p. 670.
191 the State: 395 *U.S.* 653, at 677 (1969).
191 secret law: S. Chesterfield Oppenheim, "The Patent-Antitrust Spectrum of Patent and Know-how License Limitations: Accommodation? Conflict? Or Antitrust Supremacy?" *Les Nouvelles* 6 (May 1971), p. 84 (Journal of the Licensing Executives Society, 1225 Elbur Avenue, Cleveland, Ohio 44107).
191 of preemption: Kewanee Oil Company v. Bicron Corporation et al., 478 *F. 2d* 1074 (1973).
191 Patent Laws: Ibid., p. 1078.
192 United States: Ibid., p. 1087.
192 were filed: Kewanee Oil Co. v. Bicron Corp. et al., 416 *U.S.* 470 (1974).
192 Court chamber: Ibid., p. 474.
192 the Nation: Ibid., pp. 484, 493.
193 it was: Ibid., p. 495.
193 contractual obligations: Aronson v. Quick Point Pencil Co., 440 *U.S.* 257 (1979).
194 the contract: Ibid., p. 260.
194 competition laws: Quick Point Pencil v. Aronson, 567 *F. 2d* 757, at 760 (1977).
194 trade secrets: 440 *U.S.* 257, at 263, 264, 266 (1979).
195 in decades: Diamond, Commissioner of Patents and Trademarks v. Chakrabarty, 447 *U.S.* 303 (1980).
195 of petroleum: Ibid., p. 305.
195 this title: Ibid., p. 306. Also, see *U.S. Code Annotated*, Title 35, Section 101.
195 within § 101: Application of Bergy, Coats and Malik; Application of Chakrabarty, 596 *F. 2d* 952, at 973 (1979).
195 by Congress: Parker, Acting Commissioner of Patents and Trademarks v. Flook, 437 *U.S.* 584, at 596 (1978). Also, see Thomas D. Kiley, "Common Sense and the Uncommon Bacterium—Is 'Life' Patentable?" *JPOS* 60 (July 1978), p. 468.
196 such protection: 447 *U.S.* 303, at 311, 314.
196 of innovation: Ibid., pp. 308, 309.
196 and research: Ibid., pp. 309, 310, 313.
196 it wishes: 447 *U.S.* 303, at 316-318 (1980).

197 patented process: Dawson Chemical Co. v. Rohm & Haas Co., 448 *U.S.* 176 (1980).
197 to consider: Mercoid Corp. v. Mid-Continent Co., 320 *U.S.* 661, at 669 (1944).
197 of 1952: *United States Code Annotated,* Title 35, Chapter 28, Section 271.
197 and Haas: 448 *U.S.* 176, at 179 (1980).
198 go unrewarded: Ibid, p. 221.
199 innovative companies: Stephen Wermiel, "Justices Extend Patent Safeguard on Chemical Use," *The Wall Street Journal* (30 June 1980).
199 such misuse: 448 *U.S.* 176, at 240, 241 (1980).
199 and Lutton: Diamond, Commissioner of Patents and Trademarks v. Diehr et al., 450 *U.S.* 175 (1981).
199 the press: Ibid., pp. 178, 179.
199 should issue: Ibid., pp. 179–181.
200 are used: Ibid., pp. 182, 185.
200 the analysis: Ibid., p. 188.
200 all favor: 107 *U.S.* 192, at 200 (1882). See p. 42 of this book.
200 Supreme Court: Diamond, Commissioner of Patents and Trademarks v. Bradley et al., 450 *U.S.* 381 (1981). Also, see Nelson Moskowitz, "The Patentability of Software Related Inventions After Diehr and Bradley," *JPOS* 63 (April 1981), p. 222.
201 such goods: Eli Lilly and Company v. Premo Pharmaceutical Laboratories, Inc., 630 *F. 2d* 120, at 137, 138 (1980).
202 March 12, 1981: 645 *F. 2d* 1195 (1981).
202 against it: Ibid., p. 1201.
202 its machines: "News & Comment," *PTCJ* (21 November 1974), p. A1. Also, see *PTCJ* (24 April 1975), pp. A16–A18 and D1–D8; and *PTCJ* (31 July 1975), pp. A20, A21.
202 to $111.3 million: 645 *F. 2d* 1195, at 1197 (1981). Also, see 463 *F. Supp.* 983, at 985, 986 (1978); *PTCJ* (19 March 1981), p. A1.
203 antitrust laws: 645 *F. 2d* 1195, at 1204, 1206 (1981).
204 intellectual property: Donald W. Banner, "The Patent System and the Antitrust Laws: Stabilization of the Tensions?" *Antitrust Law Journal* 49, Part I (Summer 1980), p. 44. Banner quotes Earl Pollock, in "Licensing: Patents, Trademarks and Know-how," *Antitrust Law Journal* 42 (1973), p. 631.
204 at all: Ibid. Banner quotes Tom Arnold, in "Panel Discussion: Licensing, Patents, Trademarks and Know-How," *Antitrust Law Journal* 43 (1973), p. 691.
204 to innovators: Ky P. Ewing Jr., "Antitrust Enforcement and the Patent Laws," *PTCJ* (17 May 1979), pp. D1–D4.
204 than accuracy: Abbott B. Lipsky Jr., "Lipsky Speech Attacking 'Nine No-No's,' " *PTCJ* (26 November 1981), p. D1.
205 economic irrationality: Ibid., p. D4.
205 Point case: *PTCJ* (12 October 1978), p. A5.
205 question cases: *PTCJ* (30 May 1974), p. A12.
206 critical areas: *PTCJ* (3 August 1978), p. D5.
206 patent system: *PTCJ* (5 April 1979), pp. A5, A7. Introduction of legislation S. 677 and S. 678.
206 is grotesque: *PTCJ* (1 May 1980), p. A4.
206 all effective: *PTCJ* (19 June 1980), pp. A9, A10. This summarizes the typical opposition.
206 Section of Litigation: *PTCJ* (19 June 1980), pp. A9, A10. Also, see *PTCJ* (14 February 1980), p. A4.
207 or inventors: Michael W. Blommer, "Washington Letter," *APLA Bulletin* (September 1980), p. 468. Also, see *APLA Bulletin* (December 1980), p. 738.

207 nongermane amendment: Michael W. Blommer, "Washington Letter," *APLA Bulletin* (January–February 1982), p. 9.
207 Federal Circuit: *PTCJ* (17 March 1982), pp. A1, D1. This issue contains floor remarks and text of H.R. 4482. Also, see *PTCJ* (25 March 1982), p. A1; and *PTCJ* (8 April 1982), p. A1.
207 nationwide jurisdiction: *PTJC* (7 October 1982), p. 571. For a discussion of the New Court, see Jack Q. Lever, Jr., "The New Court of Appeals for the Federal Circuit (Part I)," 64 *JPOS*, p. 178; and "(Part II–Conclusion)," 64 *JPOS*, p. 243 (April and May, 1982).
208 venerable institution: Public Law 96–517, signed December 12, 1980. This is a comprehensive law having several major aspects, one being that reexamination of issued patents is now permitted under certain circumstances. Section 9 relates to computerization.
208 of examiners: *PTCJ* (12 November 1981), pp. A1, A2, D1–D3. This contains remarks of Commissioner Gerald J. Mossinghoff before the annual meeting of the American Patent Law Association on 3 November 1981. Also, see Gerald J. Mossinghoff, "American Bar Association Address," 63 *JPOS* (July 1981), p. 342.
208 larger corporations: Public Law 97–247, signed August 27, 1982. There is no maintenance fee relative to any patent the latest application for which was filed before December 12, 1980. The fees now in effect are stated in 1022(4) *Official Gazette of the United States Patent and Trademark Office*, starting at p. 59. Also see *APLA Journal* (September-October 1982), pp. 511–513.
208 into court: Public Law 97–247, §294.
208 possible before: Sigmund Timberg, "Antitrust Aspects of Patent Litigation Arbitration and Settlement," 59 *JPOS* (April 1977), p. 247 et seq.
208 data bases: *PTCJ* (26 August 1982), pp. 422–426, at 422, 423.
209 executives alike: Ibid., p. 424.
209 of patentability: For related material, see Richard L. Gausewitz, "Brief in Support of Proposed Amendment to Section 103, Title 35, Patents, U.S. Code," *JPOS* 51 (May 1969), pp. 290–333.
210 primary status: See, for related material, Nelson H. Shapiro, "Toward a Realistic Standard of Patentability," *Idea* 16 (Winter 1974), pp. 3–13.
210 was invalid: General Motors v. Toyota, 467 *F. Supp.* 1142 (1979).
210 a decision: *PTCJ* (17 December 1981), pp. A1, A2. Also, see General Motors v. Toyota Motor Company, Ltd., et al., 212 *USPQ* 659 (1981).
210 prior art: *United States Code Annotated,* Title 35, Patents, Section 102. This section is highly complex in its various ramifications. No person should read it and think he knows what it means. There must be consultation with a patent lawyer. Such a lawyer also would explain that this section does *not* apply to the crucial field of foreign patent rights; what is permitted relative to United States patents is usually not permitted relative to foreign patents.
211 Patent Office: C. Marshall Dann, "Examination of Claims for Patentability Under 35 U.S.C. 103," *Official Gazette of the United States Patent and Trademark Office* (3 August 1976), p. 3.
212 the judge: "The Seventh Annual Judicial Conference of the United States Court of Customs and Patent Appeals," held 11 April, 1980, 88 *Federal Rules Decisions* 369, at 383, 384.
212 the night?: Abraham L. Marovitz, "Patent Cases in the District Courts—Who Should Hear Them?" *JPOS* 58 (December 1976), p. 761.
212 patent-ignorant judges: Ibid., pp. 761, 764–766.
212 patent lawyers: For biographies of all the judges on the Court of Customs and Patent Appeals, see Giles S. Rich, *A Brief History of the United States Court of*

Customs and Patent Appeals (Washington, D.C.: U.S. Government Printing Office, 1980).

213 metropolitan area: For a related plan, see Richard L. Gausewitz, "Toward Patent-Experienced Judges," *Journal of the American Bar Association* 58 (October 1972), pp. 1087–1090.

213 a whole: Marovitz, p. 767.

214 opposing side: "The Sixth Annual Judicial Conference of the United States Court of Customs and Patent Appeals," held 9 May 1979, 84 *Federal Rules Decisions* 429, at 441.

214 other kind: District Judge Frank J. McGarr, quoted in 88 *Federal Rules Decisions* 369, at 380, 381.

215 and export: Jacob Rabinow, "The Inventor and Patent Law Revision," *APLA Bulletin* (January–February 1977), p. 116.

INDEX

"A & P case," 36–42, 44
  Black Rock case and, 118–119
  Sakraida v. Ag Pro and, 144–147
Adams, Bert Nemo, 54–64
Adams battery, *see* "Neutro Cell"
Afterthoughts, 84
Aggregation-combination, 32–35, 37–38
Anderson's Black Rock v. Pavement Salvage, 115–118
Anthony, Herman, 50
"Anticipation" by earlier patent, 14
Antitrust Division, Department of Justice, 148, 203–205
Antitrust, private, 157, 202–203
Antitrust violations, 155–159
Armstrong, Major Edwin Howard, 75–76, 78–79, 87
Arnold, Thurman, 159, 167–168, 176, 203
Arnold, Tom, 204
Aronson v. Quick Point Pencil, 193–194
Atlantic Works v. Brady, 42

Ballas, George, 187
Banner, Donald W., 186
Battelle Memorial Institute, 163
Bennett, Henry, 122–124
Binks v. Ransburg, 98–100
Black, Justice Hugo L., 176–183
  preemption and, 190–191, 193
Blackmun, Justice Harry A., 73, 74, 198, 200
Black Rock case, *see* Anderson's Black Rock v. Pavement Salvage
Blenko, Walter, Jr., 106, 111
Brennan, Justice William J., Jr., 144, 192, 196, 199, 200
Brinkley, David, 186
Bruchhausen, Judge Walter, 84, 85
Burger, Chief Justice Warren E., 142, 192, 196
Bush, Dr. Vannevar, 159, 170

Calmar v. Cook Chemical, 66–69, 174
Carlson, Chester, 90, 162–163, 203
Catchwords, 30–42, 209
  aggregation-combination, 32–35, 37–38
  synergism, 118–119, 140, 145–147, 207
  whole must exceed sum of parts, 38–39, 118–119, 207
"Chakrabarty" case, *see* Diamond v. Chakrabarty
Change in degree, 29, 141
Circuits, definition of, 21
Claims, 11, 14, 133
Clark, Justice Tom C., 63, 65
Close (hard to decide) cases, 49
"Coal Stove" case, *see* Hailes v. Van Wormer
Coe, Commissioner Conway P., 167–168
Coffman, Barton, 80
Combination of old elements, 44, 65, 117–118, 136–137, 139, 207, 209
Companies, 168–170
  creative v. uncreative, 169, 170
  growth of, 89, 170
"Compco" case, 189
Computer programs, patentability of, 146, 199–200
Conflicting doctrines, 24, 26–27, 207
Conner, Judge William C., 212
"Constitutional standard," 180–182
Contributory infringement, 197
Cottrell, Frederick G., 90–91
Court of Appeals for the Federal Circuit, *see* Single court of patent appeals
Court of Claims, 60, 205
Court of Customs and Patent Appeals, 15, 205
Cronkite, Walter, 121–122
Crosby, Murray, 74–88
Crosby-Teletronics, 75–88
Crouse, William W., 95
Cuno v. Automatic Devices, 45
Cybotech Company, 103

Dann, Commissioner of Patents & Trademarks v. Johnston, 146
Darrah, William, 99, 102
Dawson Chemical Co. v. Rohm and Haas Co., 197–199
Declaratory judgment action, 17
Densmore, James, 164–165
"Designing around," defined, 160–161
De Simone, Daniel, 168
Diamond, Commissioner of Patents & Trademarks v. Bradley, 200
Diamond, Commissioner of Patents & Trademarks v. Chakrabarty, 195–197

Diamond, Commissioner of Patents & Trademarks v. Diehr and Lutton, 199–200
Differences, between invention being considered and prior art, 69–70, 171–172, 209
  size of differences, 171–172
  type of differences, 171–172
"Discovery" phase of litigation, 18, 213–214
"Doorknob" case, *see* Hotchkiss v. Greenwood
Douglas, Justice William O., 45–47, 117–118, 176–183, 190, 193, 197

Edison, Thomas, 29, 113, 171
Eibel v. Paper Co., 141
Elements, 14
Erteszek, Jan, 2–22
Erteszek, Olga, 1–22
Ewing, Ky P., Jr., 204

Federal Trade Commission, 202, 203
Federico, P. J., 43
Feldman, Leonard, 87
Ferguson, Eugene, 3
"Fixed rules," 28–30, 207
  change in degree, 29, 207
  substitution of materials, 28–29, 207
"Flash of genius" test, 45
FM radio and FM stereo radio, 74–88
Following all evidence, *see* Signposts
Fortas, Justice Abe, 64, 159
Frankfurter, Justice Felix, 45, 113–114, 177
Franklin, Benjamin, 154
Fraud on the Patent Office, 156–158
Fulton, Robert, 113

Gadgets, 174, 175, 180, 181
Galileo, 25
Gilfillan, S. C., 113, 159
Goodyear v. Ray-O-Vac, 50–51
"Graham approach," 70
Graham v. John Deere, 66–69, 84, 152, 182
Graver Tank Co. v. Linde Air Products Co., 177
Great A & P v. Supermarket Equipment Co., 36–42, 44. *See also* "A & P case"
Green, Harry, 91
Gribble, Andy, 123
Gribble, Donald Joseph, 121–124, 129–135
Griesa, Judge Thomas P., 18–20

Hailes v. Van Wormer, 32–33
Haloid Company, 163
Hand, Judge Learned, 29–30
  on combination of old elements, 44
  on Patent Act of 1952, 54
  on reaching decisions, 47–48
  on size of change in relation to prior art, 172
Hartford-Empire Co. v. U.S., 156
Hindsight, 100, 171
Holder, Judge Cale, 99
Hopkins, Samuel, 26
Hotchkiss v. Greenwood, 26–28, 45, 46
Hughes, Chief Justice Charles Evans, 46, 50
Hunt, Justice Ward, 34

Infringement (infringing), 15, 74
"Inventing around," *see* "designing around"
Inventions, how made, 23–24, 113–114, 172–173
Inventions, misunderstandings about, 171–175
Inventions, patentable and unpatentable, 70
  necessity of defining, 28
  as undefinable, 66–67

Jackson, Justice Robert H., 37–40, 44, 49–50, 178
Jefferson, Thomas, 26, 53, 151–153, 181, 196
Johnson, Merrill, 96–98
Jungerson v. Ostby and Barton, 50

Kauper, Thomas E., 205
Kewanee Oil Co. v. Bicron, 191–193
Kiker, Douglas, 186–188
Kolisch, J. Pierre, 136, 143

Land, Dr. Edwin H., 169–170
Large corporations, patent system and, 167–169
Lear v. Adkins ("Lear decision"), 190–191
"Liberty ship" case, *see* Graver v. Linde Air Products
Lincoln, Abraham, 154–155
Lincoln Engineering v. Stewart-Warner, 147
Lipsky, Abbott B., Jr., 204–205
Littlefield, D. G., 30, 31
Living matter, patentability of, 195–196

Madison, James, 25
Marconi v. U.S., 45–46
Markey, Chief Judge Howard T., 213
Marovitz, Judge Abraham L., 212, 213
Marshall, Justice Thurgood, 192, 196, 199, 200
Mass media, patent cases and, 185–188
McGarr, Judge Frank J., 211–212, 214
Meador, Daniel J., 205–206
Mechanical inventions, discrimination against, 38, 40–42, 175, 207
Mercoid cases, 197
Miller, Emery, 97, 102

Misuses of patents, 155–159, 197
Monopolies, 149–155
Mossinghoff, Commissioner Gerald J., 208, 209
Motion Picture Patents Co. v. Universal Film, 155

Nelson, Justice Samuel, 27
"Neutro Cell," 55–66
Neville, Charle, 106, 108–113
Neville, Stewart, 108, 110, 113
New quality or function, 38, 40, 41, 207
"Nonobviousness," *see* "Obviousness"
Nott, Eliphalet, 30

"Obviousness," 14, 17, 46, 67, 207, 209, 211
O'Connor, Justice Sandra Day, 201
Old elements, *see* Combination of old elements

Panels of judges, definition of, 21
Patentability tests, 26, 42, 46, 207, 209
  catchwords, 30–42, 145–147, 207, 209
  differences between invention being considered and prior art, 69–70, 171–172
  "fixed rules," 28–30, 207
  flash of genius, 45, 207
  following all evidence, *see* Signposts
  new quality or function, 38, 40, 41, 207
  obviousness, 14, 17, 46, 207, 209, 211
  severe tests, 41, 209, 210
  "skilled mechanic" test, 27, 46, 67, 207
  strictness in applying, 45, 67, 209, 210
  unusual or surprising consequences, 38, 40, 41, 207
Patentable subject matter, 195
Patent Act of 1790, 25–26
Patent Act of 1793, 26
Patent Act of 1836, 26
Patent Act of 1952, 43, 46
  Section 101, 195
  Section 103 (obviousness section), 46, 69, 70, 207, 211
Patent agents, defined, 11
Patent applications, 10–15
Patent claims, 11, 14, 133
Patent examiners, 13, 14
Patent-experienced judges, need for additional, 211–213
Patent fee bills, 208
Patent laws, *see* Patent Acts
Patent laws, proposed, 209–211
Patent lawyers, defined, 11
Patent Office, 10–15
  Board of Appeals, 15
  creation of, 26
  fraud on, 156–158
  Justice Douglas and, 180
  steps in examining procedure, 13–14
Patent protection, evolution of, 24–26
Patents
  antitrust violations and, 155–159
  as beneficial monopolies, 201
  government's motivation for granting, 24–25
  misuses of, 155–159, 197
  monopolies and, 149–155
  research (in corporations) and, 210
  suppression and, 158–159
Patent search, 11–12
Patent system
  big business and, 167–171
  constitutional purpose, 10
  misunderstandings about, 159–171
  recommendations for improving, 209–214
Pavement Salvage, Inc., 112–113
  v. Anderson's Black Rock, 114–119
"Pencil-eraser" case, *see* Reckendorfer v. Faber
Pickering v. McCullough, 35, 147
Pollock, Earl, 204
Powell, Justice Lewis F., Jr., 192, 196, 199
Precedents, 27–28
Preemption, 189–190
Preuss v. General Electric, 83–87
"Prior art," 12

"Quick Point" case, *see* Aronson v. Quick Point Pencil

Rabinow, Jacob, 215
Ransburg, Edwin, 94–95
Ransburg, Harold, 89–103
Ransburg, Harper, 90, 91, 102
"Ray-O-Vac" case, *see* Goodyear v. Ray-O-Vac
Reasonableness in applying patentability tests, 210
Reckendorfer v. Faber, 34
Reclaimix Company, 109
Reilly, John A., 59, 62–63, 174–175
Rehnquist, Justice William H., 200
Research within corporations, 210
Rich, Judge Giles S., 27, 30, 35, 43, 151, 160, 181, 195, 212–213
Roberts, Justice Owen J., 156
"Rohm and Haas" case, *see* Dawson Chemical v. Rohm and Haas
Rutledge, Justice Wiley, 45, 46
Sakraida v. Ag Pro, 135–145
Schott, Lewis, 59, 60
Science, 180–181
Sears v. Stiffel ("Sears" case), 189
"Secondary considerations," *see* "Signposts"
Severe tests of patentability, 41
"Shadow patents," 41–42

"Shift in position," 83–86
Sholes, Christopher Latham, 164, 166, 171
"Signposts" for judging patentability, 47–51, 65, 67, 70–71, 87, 115, 147, 207, 210
  following all evidence, 47–51, 65, 67
Simplicity, 41, 174–175, 209
Single-court bill, 205–207
Single court of patent appeals, 205–207
Size of differences, *see* Differences
"Skilled mechanic" test, 27, 46, 67, 207
Smith v. Goodyear Dental Vulcanite, 29
Starkey, William, 94
Statute of Monopolies, 150
Stereo broadcasting, *see* FM radio
Stevens, Justice John Paul, 199, 200
Stewart, Justice Potter, 201
Stone, Chief Justice Harlan F., 45
Strictness in applying patentability tests, 45, 67, 210
Strong, Justice William, 33
Substitution of materials, 27–30, 152
Subtests of patentability, *see* "Signposts"
Summary judgment, defined, 136
Suppression of patents, 158–159
Synergism, 118–119, 140, 145–147, 207

Technological subject matter, 175
Thomson, Elihu, 154
Trade secret law, 191
Turnham, Ellis, 36
Turnaround in the Antitrust Division, 203–205
Turnaround in Congress, 205–209
Turnaround in the courts, 188–203
Type of differences, *see* Differences

"Unobviousness," *see* "Obviousness"
Unusual or surprising consequences, 38, 40, 41, 20
U.S. v. Adams, 59–64

Vanity Fair Mills v. Olga Co., 17–22
Volta, Count Alessandro, 55

Walker, J. Bennett, 110–111
Walker Process v. Food Machinery and Chemical Corp., 156–157
Warren, Chief Justice Earl, 190
Washington, George, 88, 105
Watt, James, 113
Weinfeld, Judge Edward, 213–214
White, Justice Byron R., 64, 196, 199
Whole must exceed sum of parts, 38–39, 41, 118–119, 207
Wilson, Joseph R., Jr., 163
"Worthy patents," 41–42

Xerox Corporation, 163, 202, 203